Springerbriefs in Ecology

More information about this series at http://www.springer.com/series/10157

Jeffrey K. Keller · Charles R. Smith

Improving GIS-based
Wildlife-Habitat Analysis

 Springer

Jeffrey K. Keller
Habitat by Design
Pipersville, PA
USA

Charles R. Smith
Cornell University
Ithaca, NY
USA

ISSN 2192-4759   ISSN 2192-4767 (electronic)
ISBN 978-3-319-09607-0   ISBN 978-3-319-09608-7 (eBook)
DOI 10.1007/978-3-319-09608-7

Library of Congress Control Number: 2014947661

Springer Cham Heidelberg New York Dordrecht London

Printed on acid-free paper

Springer is part of Springer Science+Business Media (www.springer.com)

Preface

As a graduate student in the late 1970s, I became interested in causes of diversity in birds. At that time, explanations of diversity as a function of plant community structure were almost entirely based on vertical heterogeneity or complexity, usually as measured by MacArthur's foliage height diversity (FHD). However, FHD didn't work well within a given successional stage due to variations in horizontal heterogeneity among plant communities of similar ages such as successional old-fields with woody invaders. Meanwhile, measuring horizontal heterogeneity was limited primarily to on-the-ground type measures such as the coefficient of variation (CV) of distance derived from point-quarter samples, which provided only a limited amount of information on spatial arrangement of landscape elements such as shrubs and trees. I began looking for an alternative way to measure horizontal heterogeneity and fortunately discussed this with fellow graduate student Doug Heimbuch, a wizard in statistics and thinking outside the box. Together we developed what turned out to be a hexagonal-celled (raster-based) Geographic Information System (GIS) and landscape metrics package (the terms GIS and landscape metrics didn't exist, yet) that Doug facetiously dubbed spatial distribution (SPADIST). Using SPADIST, I applied some previously unexplored explanatory variables to GIS maps derived from high-resolution stereographic aerial photographs and was able to predict with a good degree of accuracy species richness and density of bird assemblages across a range of successional stages.

Although I've worked as a restoration ecologist in the private sector since that time, I have observed from a distance the development of GIS-based habitat analysis over the last several decades and noticed that most studies I read had relatively low explanatory or predictive capability. As a result, I began to look for shared, potentially limiting attributes among such studies. This book is the result of that examination and was completed with the substantial help of Charles R. Smith, another fellow grad student, mentor, sounding board, and friend of more than 35 years.

Our interpretation of what has transpired in the field of GIS-based habitat analysis is that the phenomenal technology may sometimes lead us to overlook the underlying biology. GIS provides a powerful tool for the investigation of species-habitat relationships and the development of wildlife management and conservation programs. However, the relative ease of data manipulation and analysis using GIS, associated landscape metrics packages, and sophisticated statistical tests may sometimes cause investigators to overlook important species-habitat functional relationships. Additionally, underlying assumptions of the study design or technology may have unrecognized consequences. Here we examine how initial researcher choices of image resolution, scale(s) of analysis, response and explanatory variables, and location and area of samples can influence analysis results, interpretation, predictive capability, and study-derived management prescriptions.

We begin by reviewing several remote sensing and ecological terms and discuss two shared terms with different meanings within their respective disciplines. We also revisit several longstanding concepts in landscape ecology and in some cases offer alternate points of view on their interpretation. Hopefully, these discussions are informative, but at a minimum should provide food for thought. Our aim is to urge wildlife and conservation biologists to gain a better understanding of both the capabilities and limitations of the technologies they employ and to think critically about the assumptions that underlie some of the methods and technologies, particularly those associated with GIS, now routinely applied in both research and management within these disciplines. More predictive, easily interpreted models provide a firmer basis for conservation and management decisionmaking.

We thank Steve DeGloria, Mike Morrison, Mike Scott, Angela Fuller, Amielle DeWan, Scott Stoleson, and two anonymous reviewers for providing useful comments on earlier drafts of the manuscript. An additional thanks to Mike Morrison and one of the anonymous reviewers for providing a number of recent references that helped me catch up on the literature in this rapidly evolving field. And lastly, a special thanks to Doug Heimbuch for many hours of collaboration on the original SPADIST GIS package and for elucidating several of the sampling issues addressed herein. Partial funding to C.R. Smith was provided from Hatch project NYC-147438 to Cornell University.

Jeff Keller

Contents

Introduction

Geographic Information Systems (GIS) have advanced species-habitat analyses by allowing investigators to develop an entire class of landscape metrics (e.g., McGarigal et al. 2002) that can complement traditional on-the-ground habitat quantification methods (e.g., Cottam and Curtis 1957; MacArthur and MacArthur 1961; Karr 1968; James and Shugart 1970; Tomoff 1974; Wiens 1974; Roth 1976; Aber 1979). Typical GIS programs, such as ArcGIS 10.0 (ESRI 2010), provide the analytical capacity to estimate quickly basic landscape metrics such as patch area, patch buffers, and distance between patches. More sophisticated spatial analysis software (e.g., Spatial Analyst: ESRI ArcGIS 9.0, FRAGSTATS: McGarigal et al. 2002) can produce hundreds of landscape covariates at multiple spatial scales, with varying degrees of complexity and interpretability. For example, metrics like slope gradient and aspect are easily interpreted in a biologically meaningful way. However, it may be far more challenging to understand the ecological principles underlying model-based indices such as the "Mass Fractal Dimension" derivable in FRAGSTATS (Wiens 2002). As noted by Neel et al. (2004:453–454), *"While landscape metrics have proven useful for describing landscape structure and hold promise for broader application, they are often difficult to interpret"* (also see Stauffer 2002). As a result, many studies have noted the difficulty of linking landscape pattern to species use of the landscape using these metrics (e.g., Schumaker 1996; Moilanen and Nieminen 2002; Young and Hutto 2002).

In addition, a burgeoning array of sophisticated statistical analyses (Correspondence Analysis: Hill and Gauch 1980; Canonical Correspondence Analysis: Jongman et al. 1987; ter Braak and Prentice 1988; Occupancy Modeling: MacKenzie et al. 2002; Hierarchical Modeling: Royle and Dorazio 2008) and associated software (CANOCO: ter Braak and Smilauer 1998; PRESENCE: Hines 2006; WinBugs: Spiegelhalter et al. 1999) has given researchers powerful tools to explore the significance of relationships between species or higher organizational levels and these potential explanatory landscape covariates. However, as ruefully noted by O'Conner (2002:26), *"Continued development of rigorous statistical approaches to analyzing habitat data, assisted by easy computation in the form of computing power and of packaged statistical analysis, has been unaccompanied, even to this day, by corresponding development of rigorous logic.*" (for discussion also see De Knegt et al. 2010). In a somewhat gentler comment, Morrison et al. (1998) cautioned that if our fundamental approaches to modeling are flawed, the tools used to develop models are of little importance. Thus, the ability to quantify landscapes extensively and analyze them in myriad ways does not, by itself, ensure biologically interpretable, meaningful, or managerially useful results.

In a broader discussion, Huston (2002) identified what he described as three primary impediments to developing a coherent conceptual framework for eco-logical predictions. Within the scope of our discussion of landscape metrics, these issues can be summarized as (1) mismatches between the scale of ecological measurement and the scale at which species select and use habitat, (2) misunder-standing of ecological processes such as habitat selection, and (3) use of inappro-priate statistics to quantify ecological pattern.

In response to the concerns raised above, we examine various elements of GIS-based habitat analysis for their influence on analysis results, interpretation, predic-tive capability, and development of management prescriptions and/or conservation strategies. We begin by reviewing several fundamental remote sensing and GIS definitions, two overlapping remote sensing and ecological terms, and several additional terms relevant to GIS applications in ecology and landscape ecology. While reviewing these latter terms, we argue that the presence and abundance of individual species can be linked in GIS to "patches" of taxon-scaled and -specific landscape components, classifiable as either "solid" (e.g. open grass, water, decid-uous trees) or "edge" (e.g., shrub-grass, emergent marsh-open water) types on remotely sensed imagery.

In subsequent chapters, we explore application of this concept in GIS-based habitat analysis. We review current analytical approaches and scales of analy-sis, and compare these to alternatives where (1) image resolution and selection of explanatory variables are more closely scaled to the body size and landscape use of the focal taxon, (2) variable selection considers theoretically optimal patch shape based on energetics, and/or (3) the location and scale of habitat samples more closely match species detection locations and scales of habitat use. Lastly, we demonstrate the strength of an analytical approach that employs the preced-ing three elements by comparing the explanatory power of four commonly used landscape metrics with a set of more species-scaled independent variables in describing the habitat relationships of seven species of breeding birds. Although our discussion will primarily reference birds and other terrestrial vertebrates, the principles of scale and variable selection described can be applied by wildlife researchers, managers and conservation biologists to aquatic and invertebrate taxa as well.

Keywords

GIS · Habitat analysis · Patch shape · Edge · Image resolution · Landscape metrics · Remote sensing · Scale

Chapter 1
Working Definitions

1.1 Remote Sensing and GIS

A thorough discussion of remote sensing and GIS terminology is beyond the scope of this paper, and the reader is directed to Lillesand et al. (2008) and Campbell and Wynne (2011) for more in-depth treatment of these subjects. However, several concepts and definitions within this realm are critical to understanding the influence of remotely sensed image selection on (1) landscape component identification and (2) resulting GIS-based analyses, interpretations, and predictive capabilities of those analyses. Some relevant definitions follow:

Remote Sensing is the use of aerospace sensor technologies to detect and classify objects on earth (on the surface, in the atmosphere, and oceans) by means of passive remote sensors, such as photographs detecting reflected sunlight, or propagated signals, including electromagnetic radiation such as radar or LiDAR (see below), emitted from aircraft or satellites. Remote sensing data are often used to provide the principal data source in a GIS, and their attributes should be carefully considered, as we shall discuss.

GIS is the acronym for geographic information systems. A GIS integrates hardware, software, and data for capturing, managing, analyzing, and displaying all forms of geographically referenced information (http://www.esri.com/what-is-gis/overview.html). A GIS typically uses remote sensing data as the underlying source of information to be manipulated and/or analyzed with the GIS's associated software.

Pixel In remote sensing, it is the smallest single component in a digital image. The address of a pixel (picture element) corresponds to its physical location. Also, historically, the smallest unit of information in a raster-based GIS format (i.e., a cell or grid cell), the size of which if larger than a true pixel, is user defined.

© The Author(s) 2014
J.K. Keller and C.R. Smith, *Improving GIS-based Wildlife-Habitat Analysis*,
SpringerBriefs in Ecology, DOI 10.1007/978-3-319-09608-7_1

In either case, true pixel or grid cell, this single smallest element is, by definition, classified as a single type within whatever classification system is applied. However, it may actually contain multiple landscape components within it when viewed at the *resolution* to which the organism of interest responds, a point rarely mentioned, and by inference, apparently rarely considered a priori, in GIS-based habitat studies.

Minimum Size Delineation (MSD) is the minimum polygon size used to delimit a single feature of interest (see also minimum mapping unit).

Minimum Mapping Unit (MMU) is the minimum size of a particular class type within a classification system (e.g., spruce-fir cover type; Parker soil series). It is often used interchangeably in ecological applications with minimum size delineation.

Image Resolution The standard four components of image resolution are *spectral* resolution (the number, width, and location of wavebands that compose a digital image), *spatial* resolution (smallest resolvable unit, grd, gsd, ifov, see below), *temporal* resolution (frequency of acquisition, repeat cycle, seasonal), and *radiometric* resolution (number of bits, 2^n, per waveband used to capture the spectral variation in a remotely sensed scene; most digital imagery now ranges from $n = 8$ or 256 gray levels to $n = 11$ or 2,048 gray levels). *Thematic* resolution can be added as a fifth component. It refers to the level of categorical detail the user wishes to extract from an image. For example, if one wishes to map only water, high spectral, spatial, temporal, or radiometric resolution is not required. However, if one wishes to map a small ephemeral feature such as a vernal pool that exhibits variable spectral response through time and map it at a ground resolvable distance (GRD) of less than ~1 m, then imagery of high spectral, spatial, temporal, and radiometric resolution is required. Additionally, one must consider related factors such as "image contrast," which is a function of spectral response between the feature of interest and background (e.g., a white golf ball on green grass is more detectable than a green golf ball on green grass). Of these resolution components, *spatial resolution*, measured as GRD, is particularly relevant for our discussion but should not be confused with *pixel size*, with which it is sometimes equated in ecological contexts (e.g., Ostapowicz et al. 2013: 1108).

Ground Resolvable Distance (GRD) is the minimum distance between two objects that allows the two objects to be identified as separate entities. Assessed by image analysts, it directly indicates the resolving power of an imaging system, a critical consideration in habitat modeling where the ability to separate habitat from the surrounding matrix depends on the ability to accurately identify and quantify the spatial attributes of individual landscape components used in habitat selection by the species of interest. Despite GRD's importance as the most direct gauge of an image's appropriateness for a particular study, only a small fraction of the GIS-based habitat analyses we reviewed included the value of this standard measure of image quality for the selected data source.

Ground Sampled Distance (GSD) is an often used alternative to GRD; it is the center-to-center distance between adjacent pixels in an image. However, unlike GRD, it is not a measure of image quality and refers only to the detector sampling projected onto the ground. Also, unlike GRD, it does not account for effects that the optical system, atmosphere, or target movement may have on image resolution.

Instantaneous Field of View (IFOV) is a measure of the ground area viewed by a single detector element of a sensor at a given instant in time. On a resulting image, it is the pixel width and represents the lower limit of object size that can be resolved in the image (i.e., the smallest resolvable object cannot be smaller than the pixel size). The IFOV is determined by the focal length of the sensor and its height above the ground. The width of the ground instantaneous field of view (GIFOV) is equivalent to the GSD. Like GSD, it is not a measure of image quality, only an indicator of the upper limit of image resolution.

National Imagery Interpretability Rating Scale (NIIRS) is an expert-based ranking system of image quality based on interpretability. It "defines different levels of interpretability by the types of tasks an analyst can perform with imagery of a given rating level" (Encyclopedia of Optical Engineering 2003). First developed for military use, the NIIRS now includes gradations based on civilian features (Appendix A). It is of particular interest to non-remote sensing specialists because it provides a powerful, practical method to assess an image's resolution, a consideration critical to ecologists using geospatial tools to investigate species–habitat relationships. However, perhaps due to its relatively recent development, no studies we reviewed referenced this standard.

Image Classification is the process of reducing a remotely sensed image to information classes. Regardless of the approach applied to image classification (e.g., supervised or unsupervised classification of digital imagery versus manual classification of individual cells or polygons in aerial photography), both the level of detail in a classification system and the accuracy of a classification employing that system are functions of the underlying image resolution (GRD). Thus, particularly for smaller species, the ability to accurately classify individual landscape components such as trees, shrubs, small bodies of water (e.g., vernal pools), etc. at a resolution relevant to an individual's territory-scale use of the landscape is often essential to accurately characterize and quantify habitat relationships in a biologically meaningful way (Wiens et al. 1987).

Image or Mapping Scale is the ratio of the distance on an image to the corresponding distance on the ground, measured in the same units (e.g., if 1 cm on the image = 40,000 cm on the ground, the resulting image scale is 1:40000).

LiDAR light detection and ranging is an optical remote sensing technology that can measure the distance to, or other properties of a target by illuminating the target with light, often using pulses from a laser. By recording different types of backscattering of light from LiDAR laser pulses, it is possible to map various features of interest based on wavelength-dependent changes in the intensity of the returned signal. In ecology, LiDAR is used to measure canopy height, biomass, and leaf area and its vertical distribution, which, in turn, has been correlated with the distribution of taxa dependent on specific distributions of leaves at different levels within a forest (e.g., Goetz et al. 2010).

1.2 Remote Sensing and Ecology: Shared Terminology

1.2.1 Resolution

Kotliar and Wiens (1990), using Wiens (1976) definition of patch as "a surface area differing from its surroundings," suggested that the finest and coarsest scales at which an organism responds to patch structure by differentiating among patches could be defined as grain and extent, respectively. They noted that at a scale smaller than the grain, the organism functionally perceives its environment as homogeneous (Kolasa 1989). Wiens (1989a: 388) also observed that *"as grain increases, a greater proportion of the spatial heterogeneity of the system is contained within a sample or grain and is lost to our resolution."* (see definition of Pixel above).

For someone studying species–habitat relationships using GIS, the ecological term *grain* is thus analogous to the remote sensing definition of *spatial resolution*. This is because in a GIS, grain for an organism becomes the smallest landscape component that is both discernible in the image as defined by GRD and used by the organism in habitat selection. Consequently, the ability of the image to resolve objects at least as fine as the grain may be critical.

Kotliar and Wiens' term *extent*, though defined as the coarsest **scale** (our emphasis) of organism response to the heterogeneity of its environment, actually represents a geographic **area**, defined as equivalent to the lifetime home range of the organism. This organism-defined maximum area is typically unknown, but has come to be user defined in multi-scale GIS-based analyses as the largest of the nested set of areas analyzed in such studies (e.g., Howell et al. 2008). Similarly, all more fine-grained *scales* (see definition below) examined within a multi-scale hierarchy also are actually sampling *areas* within which the researcher attempts to identify patterns of organism use related to habitat attributes such as vegetation composition and component spatial arrangement.

Regardless of the size of the area analyzed in a single-scale study or the range of area sizes analyzed in multi-scale studies, the *resolution* of the imagery employed is fixed at the outset by the chosen remotely sensed data source. Note, for example, how in Fig. 1 of Wiens (1989a), which is the equivalent of a remotely sensed image, only the proposed sampling areas representing grain and extent change, not the patches or their spatial arrangement. Thus, in a GIS, it is the choice of image resolution that dictates the minimum size of individual landscape components (i.e., the absolute level of grain) that can be identified and quantified. Since grain should be organism-based (Kotliar and Wiens 1990), this choice is critical in any such study. Yet, few studies appear to recognize a priori the full implications of the choice of image spatial resolution (and associated classification system) on GIS-based analyses, results, and study inferences (McDermid et al. 2009, 2010). We detail these issues in Chap. 2 (also see Arponen et al. 2012).

1.2.2 Scale

1.2.2.1 Discipline Differences

Interestingly, the introduction of remotely sensed imagery to ecology via GIS-based habitat analysis has juxtaposed two preexisting, and essentially, opposite interpretations of the land area in question when using the term "scale." Cartographers refer to a "small scale" (e.g., 1:250000), when speaking of a large land area (typically with a lower image resolution), while this same expression in ecology refers to a small spatial extent viewed at the level of individual landscape elements or component parts (Wiens 1989b). Conversely, large mapping scales in cartography (e.g., 1:12000) refer to a small area of land that typically is viewed at a higher resolution, which provides the ability to resolve more components on the ground. In ecology, "large scale" refers to a large land area, typically one considered as composed of spatially more extensive units such as plant communities (also see Morrison and Hall 2002). Thus, smaller animals with typically smaller home ranges or territories actually require higher-resolution imagery in order to facilitate adequate depiction of the composition and spatial arrangement of elements composing their habitats.

1.2.2.2 Potential Confusion

In remote sensing, scale is precisely defined as the ratio of a mapped or on-image distance to the corresponding on-the-ground distance and is distinct from image resolution. In landscape ecology, the term has developed a very different and arguably confounding usage.

With regard to ecological scale, Wiens (1989a: 387) stated that *"Any inferences about scale-dependency in a system are constrained by the extent and grain of investigation…and we cannot detect any elements of patterns below the grain."* He defined grain as *"the size of the individual units of observation, the quadrats of a field ecologist…"* (i.e., an area). He defined extent as *"the overall **area** encompassed by a study"* and noted that *"extent and grain define the upper and lower limits of **resolution**"* (again, our emphasis) *of a study,"* implying that resolution changes from grain to extent, which in GIS, as noted above, it does not. However, his discussion of this topic did not reference GIS and he explained his thinking on changing resolution thusly: *"As grain (sampling area) increases, a greater proportion of the spatial heterogeneity of the system is contained within a sample or grain and is lost to our resolution."* (Wiens op. cit.: 388). He further explained using Figs. 1 and 2 (op. cit.) that small areas contain less variance associated with measurement of landscape component arrangement, thus equating smaller sampling areas with increased resolution of attributes potentially associated with a species occurrence. In GIS, actual measurement of such attributes and associated variance of those measures will be a function of the image resolution and classification system employed.

Although Wien's discussion of scale derives from ground-based sampling, by the above definitions, grain and extent when applied in a GIS context are simultaneously resolution *and* a sampling area, thus equating the term *scale* with both resolution and area, a concept that is at odds with scale as applied in remote sensing. Kotliar and Wiens (1990) appeared to remove some of the definitional confusion of grain and extent as being simultaneously area and resolution by defining them in terms of the focal taxon as the finest and coarsest scales at which an organism responds to patch structure by differentiating among patches. Even under this somewhat less confounding definition, Huston (2002) observed that ecological scale considers two independent (i.e., measured using different units) and unrelated elements (grain, defined as resolution, and extent, defined as total area) as endpoints along a single continuum.

For GIS-based studies, this interpretational inconsistency has remained. Wiens later (2002:747) again equated areas of analysis with resolution by referencing "*scales of resolution*" and using the example of "*several discrete window sizes*" (i.e., areas). Similarly, Morrison and Hall (2002: 44), in a reference to the term "extent," suggested that ecological scale is "*the **resolution** (again, our emphasis) at which patterns are measured, perceived, or represented*" (i.e., that both grain and extent represent resolution). They further defined *grain* as "*the smallest resolvable **unit of study**"* and gave the example of a 1×1 m quadrat (i.e., an area) similar to Wiens (1989a), while also continuing to define extent as an *area* per Kotliar and Wiens (1990).

In actuality, ecological scale, viewed as both an area and the resolution of measurement, as originally described in relation to on-the-ground studies (Wiens 1989a), is considered very infrequently in GIS-based analyses. Only a few of the multi-scale GIS-based studies we reviewed varied image resolution (by aggregating cells or mmu's) to test the effects of using progressively lower resolution (e.g., Thompson and McGarigal 2002). Instead, almost all multi-scale studies varied only the *size* of the *sampling area* on the fixed-resolution GIS map employed. Additionally, few of these studies considered the limitations of that resolution on landscape component identification, quantification, and ultimately the ability to link biologically measures derived at that resolution to species occurrence (but see Farrell et al. 2013). We suggest that without considering the influence of image resolution in multi-scale GIS-based analyses, either a priori or within the analysis, interpretations of the supposed influence of ecological scale may be called into question (see Sect. 2.5).

Although we shall use scale in the currently accepted ecological sense throughout our discussion, we found its application in GIS to be more constricted to the notion of changing sampling area sizes than as originally conceived and applied in the context of ground-based studies. We submit that in a GIS context, its definition is relatively imprecise (encompassing resolution and area as endpoints of the same continuum) and, given its actual application, potentially misleading, especially in light of the preexisting, mathematically more precise, and quite separate definitions of the terms scale and resolution in remote sensing.

1.2.2.3 Scale or Simply Level of Context?

As noted in the foregoing discussion, the term scale actually is applied most routinely to the range of sizes of *sampling areas* defined by the researcher in multi-scale GIS-based analyses. For example, a typical study of the influence of landscape heterogeneity on a species or assemblage might include analysis of mapped remote sensing data within three progressively larger sampling areas (e.g., a *local* scale, a *patch* scale, and a *landscape* scale) circumscribing a point where information about the focal taxon (e.g., detection/non-detection, density, reproductive success) is gathered. Each nested sample is replicated to produce a statistically viable number of samples of each of the three "scales" of analysis.

Also, as noted above, measurement of GIS-based covariates is constrained at all ecological scales (i.e., sampling area sizes) by the resolution of the original imagery, which never changes over the range of grain to extent examined by the researcher. Rather, only the *area sampled* (i.e., the ecological scale) changes. This means that at each sample area size, observed heterogeneity (measured as number or spatial arrangement of patches of a given type, range of patch sizes or shapes, edge density, etc.) and its inferred influence on the focal taxon are largely dependent on the initial choices of image resolution and classification system. Therefore, viewed simply as the changing sample areas they are, so-called ecological scales might be described more accurately as local, patch, landscape, etc.—*levels of context* within which the focal taxon is examined. Note: We use the term *level* here as defined by Maurer (2002: 126) as a theoretical construct used *to induce conceptual order to thinking* about complex systems. That is, levels are researcher defined rather than an empirical construct used *to organize data* about complex systems (op. cit.), which is the description Maurer applied to the term *scale*.

1.3 Community and Landscape Ecology

Despite the inconsistencies cited above, scales of analysis have been a topic of intense interest among researchers since the introduction of GIS to ecological studies (e.g., Urban et al. 1987; Kotliar and Wiens 1990; Brown 1995; Jelinski and Wu 1996; O'Neill et al. 1996; Collins and Glenn 1997; Gaston and Blackburn 2000; Pearman 2002; Riitters et al. 2002; Scott et al. 2002; Thompson and McGarigal 2002; Cushman and McGarigal 2004; Wu 2004; Wu and Hobbs 2006; Smith et al. 2008). Consequently, the terminology and definitions of scale-related metrics, such as edge density or habitat permeability, and associated analyses also have developed rather recently, and this has been in association with widely disparate applications. Not surprisingly, these metrics are not uniform. Additionally, despite repeated attempts at clarification/unification (e.g., Allen and Hoekstra 1992; Block and Brennan 1993; Hall et al. 1997; Morrison and Hall 2002), much ambiguity of hierarchical terms of ecological organization still is reflected in current usage (reviewed by Looijen 1998, Chap. 10). Taken together, the inconsistencies in terminology of both scale-related metrics and community ecology can result in alternate, sometimes confusing usages of terms or even unintentional misrepresentations of underlying ecology. We therefore feel it is useful to review some of this terminology.

1.3.1 Hierarchical Analytical Scales

Wiens et al. (1987) and Wiens (1989b) suggested 4 increasingly larger scales of analysis to better determine species–habitat associations: (1) the within-plot scale that references the space occupied by an individual; (2) the local-patch scale where individual patches are occupied by multiple individuals of multiple species; (3) the regional scale, which includes multiple patches occupied by many local populations; and (4) the biogeographic scale encompassing different vegetation formations, species assemblages, and climates.

The first three scales in Wiens' hierarchy appear to correspond to preexisting ecological terms and/or concepts. For example, Wiens' within-plot scale corresponds to the traditional concept of habitat (e.g., Kendeigh 1961). The local patch corresponds with community (i.e., a plant community and its associated animal community; Forman and Godron 1981) and the regional scale corresponds with the current ecological use of the term landscape (Forman and Godron 1981; Forman 1995). Other potentially useful analytical scales and associated organizational concepts also exist and include some less frequently applied constructs and terminology.

1.3.2 Habitat

At the most basic level (sensu Maurer 2002) is the individual species. Following decades of qualitative usage that often considered habitat an element of the Grinellian niche (i.e., niche = habitat + niche; cf. Grinnell 1917, 1924, 1928), Hutchinson (1957) attempted to describe the niche more quantitatively as a multi-dimensional space defined by a species' collective frequency distributions across a set of biotic and abiotic variables (resources). Whittaker et al. (1973) argued that this view focused on the *within community* position of a species, thus conceptualizing niche more per the Eltonian concept as the *functional role* of a species (Elton 1927; reviewed by Wiens 1989b, Morrison and Hall 2002; Soberon 2007).

On the heels of Hutchinson's (1957) attempt to refine the concept of niche, other authors, following separate historical usage (see references in Whittaker et al. 1973), offered generally non-overlapping (with niche), species-centric definitions of habitat as the *standing place* or *living place* of an organism (Udvardy 1959; Kendeigh 1961; Hanson 1962; Odum 1971), as separate from its function. Whittaker et al. (1973) argued for and provided additional quantitative definitions for both the place (habitat in the sense of variable space, not actual location space) and function (niche) of a species in the landscape. Implicit in their discussion and those of the preceding authors was the correctness of the oft used lay definition of habitat as "the place where a species lives," described by (the values of) a combination of biotic and abiotic variables (i.e., habitat as a subset of an environment sensu Looijen 1998: 165; also see Keller and Smith 1983).

Block and Brennan (1993) reviewed the history of the habitat concept from an ornithological viewpoint, and Hall et al. (1997) reviewed use/misuse of the term and proposed standard definitions for it and related terms. Lastly, Looijen (1998) and Morrison and Hall (2002) provided a thorough discussion of the history of various ecological terms and argued for further refinement in their usage to achieve desirably unambiguous and mutually exclusive definitions. These authors continued to maintain, however, the connection between an organism and habitat as a "place" defined by biotic and abiotic descriptors, although in the case of Whittaker et al. and Looijen, not necessarily in the sense of a particular physical location.

It then follows that although habitat is a fundamental level of hierarchical organization, its characterization depends on identification of variables that describe attributes (e.g., height, diameter, density, depth, temperature, pH, aspect, slope, etc.) of its component parts (e.g., trees, shrubs, soil, water, topography, etc.) as well as of the species' habitat as a whole (e.g., habitat size, insularity, configuration). At higher organizational levels (see below), these same variables plus additional, more spatially extensive attributes are required to characterize adequately these broader entities. It is at this organizational or geographic scale interface that definitional separation in variable description (e.g., habitat vs. biotope vs. landscape) has routinely been dropped or overlooked for decades, either viewing all these attributes as simply "habitat variables" or redefining habitat, often anthropogenically rather than organismally, as "microhabitat" when referring to intraterritory, component parts, small spatial extents or small animals (e.g., Lemen and Rosenzweig 1978; Dueser and Shughart 1978; Dettmers and Bart 1999; MacFaden and Capen 2002; Barg, et al. 2006), and "macrohabitat" for larger animals, spatially more extensive elements, or broader concepts (e.g., forest habitat). See Wiens and Milne (1989) for a discussion.

In many instances, this lack of precision has little or no consequence. However, when discussing the various levels in relation to one another, as is critical (1) in hypothesis formulation, (2) when comparing studies, or (3) in development of conservation plans for endangered species versus more diverse assemblages, the desirability of and utility in maintaining the definitional distinctions and precision of usage is, in our opinion, reaffirmed (cf. Franklin et al. 2002 for a similar discussion of *habitat fragmentation*).

1.3.2.1 Habitat Selection

Discussion of habitat associations readily leads to consideration of scales of habitat selection by organisms. Johnson (1980) suggested a widely recognized hierarchical course-to-fine-grained habitat selection process where first-order selection pertains to a species geographical range, second-order selection defines home range, and third-order selection refers to usage of subsets of the home range (i.e., particular habitat components). We independently described essentially the same hierarchy in a previous discussion of habitat using the red-eyed vireo (*Vireo olivaceous*) as an example (Keller and Smith 1983: 20–21). We noted, "*A quick check of its range map in one of the popular field guides shows that in summer it is distributed over much of the continental United States and Canada* [Johnson's first-order selection]. *However, we know that within that broad geographic range red-eyed vireos are not found everywhere. Typically, they inhabit deciduous forests* [Johnson's second-order selection]....*Within the deciduous forest community, red-eyed vireos usually inhabit the tree canopy* [Johnson's third-order selection]. *Thus, exactly where a bird lives turns out to be much more restricted than a range map implies.*" We later continued "*...the red-eyed vireo makes its living by searching among the leaves and twigs of the forest canopy for insects.*" This represents a fourth-order selection process identified by Johnson (1980), in this case, procurement of food items from those available within the site identified by third-order selection. By this description, "selection" more closely approximates "habitat use" as "*the way an animal uses* [*or* "*consumes,*" *in a generic sense*] *a collection of physical and biological components* [i.e., *resources*] *in a habitat.*" (Hall et al. 1997: 175) than "habitat selection." Thus, as used here, within-component foraging equates to a species' ecological function (i.e., niche; e.g., MacArthur 1958).

Although a discussion of niche is a logical progression here and, arguably, niche variables influence a species reproductive success within its habitat, niche is not routinely quantified via remote sensing, and we direct the reader to other discussions of the subject (cf. Morrison and Hall 2002: 46). Overall, Johnson's suggested hierarchical habitat selection process has served as a template for many GIS-based multi-scale habitat selection investigations.

1.3.3 Biotope

Looijen (1998) recommended the use of the term biotope to describe the physical place of a community as "*an area* [*topographic unit*] *characterized by distinct, more or less uniform, biotic and/or abiotic conditions ...*" (p. 165 Box 5). Under this definition, biotope is essentially the equivalent of the current landscape ecology/GIS interpretation of Wiens (1976) definition of a patch as a surface area differing from its surroundings (see next section).

Looijen (1998) based his recommendation of use of the term biotope on prior definition (e.g., Udvardy 1959; Kendeigh 1961; Hanson 1962; Whittaker et al. 1973) and current usage of the term in both English and other European languages (Looijen 1998). He also noted, however, several other definitions, including biotope as the "environment" of a community (i.e., without reference to a particular physical location). This latter definition is more consistent with the definition described above of habitat, which is independent of physical location and relies only on the set of (values of) biotic and abiotic variables that characterize its occurrence. However, both concepts appear to have utility in community and landscape ecology.

A biotope's physical location is sometimes discrete enough to identify in remotely sensed imagery, particularly in landscapes that are human dominated (e.g., agricultural) or characterized by rapid hydrological or elevational transitions, or sudden geologic discontinuities. For example, the talus slopes (adjacent to deciduous forest) occupied by the Allegheny Woodrat (*Neotoma magister*) represent an abiotic land cover that typifies an appropriate use of both the "physical place" concept of biotope and the variable characterization concept. However, the boundaries of more gradual transitions (i.e., gradients) between natural communities are often better described by the combinations of biotic and abiotic variables of the latter concept (i.e., independent of physical location).

In truth, biotope appears to be largely absent from the more recent North American ecological literature. We find this unfortunate because the term groups such disparate but hierarchically similar entities as a coral reef, a stream channel, and all manner of plant communities (as containing habitats for animals) under one organizational umbrella. We include it in here for its utility in landscape ecology and GIS applications as a relatively unambiguous descriptor of the "*habitat of a community.*"

Finally, for both habitat and biotope, one may consider the potential versus realized range of values of the explanatory (biotic and/or abiotic) variables. We will limit our discussion to realized examples since most analyses deal with measuring habitats or biotopes where organisms currently exist. This is certainly true for model-building purposes.

1.3.4 Habitat in GIS

As noted in our discussion above, habitat is most typically considered a "place," either physically or in variable space, defined by (the values of) a suite of biotic and abiotic descriptors. Due to its very nature, a GIS image represents an abstraction of the landscape to a reduced set of these variables for any taxon under consideration. In GIS, whether raster based or vector based, information in the underlying remotely sensed image is categorized and analyzed in only two forms, pixels (or grid cells) and the boundaries between them. This is regardless of how many layers of information (e.g., vegetative and topographic) are included.

Pixels or grid cells are manually or through an automated process classified to type and then grouped as either clusters (raster-based) or polygons (vector) of like types. These clusters or polygons are typically referred to in landscape ecology as "patches." When landscapes are analyzed with metrics packages such as Spatial Analyst (ESRI ArcGIS 9.0) or FRAGSTATS (McGarigal et al. 2002), covariates describing the interfaces ("edges") between patches also can be calculated. Within the limitations of this abstraction of the full biotic and abiotic environment to a visual representation, how are these two major features of a GIS landscape, patch and edge, viewed and interpreted by ecologists in relation to species' use of the landscape?

1.3.4.1 Patch

Among the earliest ecological applications of the term patch, MacArthur et al. (1962) introduced the concept that many terrestrial bird species are associated with particular structural profiles of vegetation, which they termed "patches," within landscapes, regardless of the moniker (i.e., oldfield, clear-cut, mature forest, etc.) applied to the plant community where the patch occurs. They went on to demonstrate, in what was an elegant and at the time, extremely quantitative way that knowledge of avian–patch associations could be used in a predictive fashion.

Wiens (1976), connoting a plan view perspective of landscape heterogeneity in contrast to MacArthur et al.'s (1962) vertical profile concept, proposed defining patch as "*a surface area differing from its surroundings in nature and appearance.*" This definition, which lends itself readily to GIS applications, has largely been adopted as the standard interpretation of the term patch in ecology. Particularly in landscape ecology, it is most frequently used to refer to a discrete and internally homogeneous entity such as a plant community or biotope (see section above) in the landscape (Forman and Godron 1981; Wiens 1989b). Conversely, note how the current usage of *patch* as the equivalent of a plant community or biotope (i.e., a land use or cover type in a GIS) differs from Kotliar and Wiens' (1990) earlier description of patch as organism defined and at a potentially very high resolution (e.g., they use the example of a single flower as a *patch* for a nectarivore). Kotliar and Wiens also noted that "discrete" patches are rarely observed in nature. Instead, they suggested (op. cit.: 253), "*Ideally, 'patches' should be determined using objective criteria to define their boundaries* (e.g., *the relative rate of change in a variable of interest per unit of space...*", and that "*Such criteria could form the framework for comparison of patch structure between different systems or within the same system.*" We explore the application of these criteria to GIS interpretations of landscape structure in the following two sections.

1.3.4.2 Edge

The concept of "edge" is applied in various ways. In wildlife biology, it has been used since Leopold (1933) to describe vegetation associations of many game species and is defined most typically in ecology as the boundary between two plant communities. This definition, known as an ecotone (Clements 1905; Odum 1971), has become the default concept of edge in landscape ecology. This may be due, at least partially, to the fact that some of the earliest GIS applications in landscape ecology were associated with documenting the impacts of anthropogenically generated forest fragmentation on forest interior species (e.g., McGarigal and McComb 1995; Trzcinski et al. 1999 and references therein). As with many game species, most of these studies considered only ecotone-scale edge quantified within extensive landscapes using relatively low-resolution Landsat imagery or small-scale aerial photography (i.e., 1:50000). These studies and much subsequent work on fragmentation, which have linked edge with generalist species, nest predators, and nest parasites, have, in our opinion, not only fostered a negative connotation of edge among conservation biologists, but have led to some misconceptions about edge among ecologists, in general.

First, the ecotone view of edge implies that (1) there are relatively few types of edge in a landscape and (2) individually, those edges are relatively extensive spatially (cf. Fig. 1.1; e.g., Forman and Godron 1981). In contrast to this view, we agree with Risser (1987) and Wiens (1989b) that myriad edges of numerous types are present at many scales, both between and, frequently, within biotopes (Fig. 1.2). This alternate view also suggests that there are many edge types both biotic and biotic–abiotic that, viewed at the scale perceived by the organism, are recognized by various species of wildlife as appropriate habitat. Examples include the emergent marsh–open water edge used by marsh birds (Rallidae and Ardeidae) (Rehm and Baldassarre 2007), the deciduous forest–talus slope edge used by the Alleghany Woodrat, referenced earlier, and the canopy–open air edge used by various flycatchers (Fig. 1.2b).

Second, analysis of edge (1) often considers all edge to be equivalent, (2) is accomplished using low-resolution imagery, and (3) employs landscape metrics that do not adequately quantify spatial arrangement (see Chap. 3). This has resulted in poor or completely misleading correlations of edge with various taxa, particularly with smaller non-game species of conservation interest, points we discuss in greater detail in subsequent chapters.

Lastly, edge is by definition a border or linear boundary, and its appearance only in plan view on a computer screen GIS map at a typically small scale (e.g., 1:40000), as simply a line, may subtly further the notion among some GIS users that edge merely represents the demarcation between plant communities or land use types (i.e., it is a landscape feature without area). Contrary to this potential misinterpretation, the historic value of edge for wildlife was considered to lie in the *composition* of its adjacent landscape components (e.g., one component used for foraging, the other for nesting and escape cover), not simply the linear interface between them (Leopold 1933).

1.3.4.3   Edge as a Form of Patch

Considering edge only at an ecotone scale as a 1-dimensional boundary on a GIS map and typically quantifying only its length or contrast without regard to actual composition (e.g., McGarigal and McComb 1995; Saab 1999; Brennan and Schnell 2007) misses the importance of (1) the landscape elements composing the boundary (Keller and Anderson 1992), (2) the spatial extent (2-dimensionality) of those adjacent components in a GIS, and (3) their actual 3-dimensional nature on

Fig. 1.1 A 1:40000 view of the Connecticut Hill WMA in the southern Finger Lakes region of central New York. *Source* Google Earth May 1, 2007, New York GIS

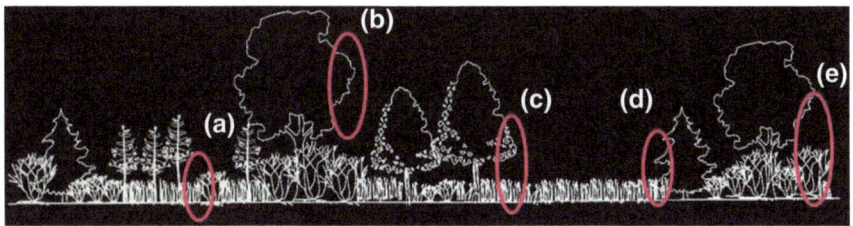

Fig. 1.2 A perspective view of a mid-stage successional oldfield illustrating 5 different types of edge to which various species of wildlife respond. *a* shrub/grass, *b* deciduous canopy/open air, *c* deciduous sapling-poletimber/grass, *d* coniferous sapling-poletimber/grass, *e* deciduous canopy/ deciduous shrub. Many studies aggregate all these types together as simply "edge," view them only as "contrasts," or do not even consider them because they are unidentifiable when using lower-resolution imagery

the ground. It is the *composition* of any edge at a *taxon-specific resolution* and the *spatial arrangement* of that edge composition on the landscape (see Sect. 3.2.3) that determines a particular species' response to it as habitat or non-habitat. When viewed in this context as a potentially inhabitable space (i.e., a territory), edge indeed occupies an *area* on a GIS map (i.e., the edge and elements composing it are 2-dimensional) in the same way that polygons or cell clusters of a particular landscape component occupy an area (e.g., clusters of trees = forest). As such, edges composed of particular pairings of landscape components simply represent an alternate form of "patch" with which various "edge" species may be associated in the same way in which other species are associated with polygons or cell clusters of particular component types (e.g., red-eyed vireo with forest).

We submit that this interpretation of edge, as simply another form of patch (i.e., having area), is entirely consistent with Wien's (1976) definition of a patch as "*a surface area differing from its surroundings in nature and appearance*" and discuss its quantification as such in Chap. 3. Therefore, from a GIS perspective, we additionally suggest the following definitions to separate what we consider to be two major patch associations of wildlife discernible from remotely sensed imagery (Keller 1986, 1990; Keller et al. 2003):

SOLID—*clusters* (a polygon in vector-based GIS, or contiguous pixels or grid cells in raster-based GIS; see Keller et al. 1979a; Turner 1989) of identical or structurally similar landscape components (e.g., open grass, open water, sawtimber trees [i.e., a forest], emergent marsh) associated with a particular species or assemblage.

EDGE—any combination of *interfaces* between adjacent structurally dissimilar landscape components (e.g., shrub–grass, deciduous tree–grass, open water–emergent marsh) associated with a particular species or assemblage.

These terms are in many ways an extension in plan view of MacArthur's patch type and vegetation profile concepts, as well as James' (1971) gestalt concept, as general constructs that help explain habitat associations and the individual distribution of species across landscapes regardless of any labels (e.g., forest-field edge, oldfield, clear-cut) used to identify the locale (cf. Bullock and Buehler 2006; King et al. 2009). As one example, using the forest/field ecotone and the two biotopes (oldfield, clear-cut) just mentioned, the common yellowthroat (*Geothlypis trichas*) occurs in increasing densities from the forest/field ecotone through the two biotopes, not because of the type of ecotone or plant community per se, but because the dense, low vegetation it favors, composed typically of stump sprouts, root suckers, or shrubs, occurs with increasing abundance (threshold level through optimal habitat) across the three landscapes (Keller 1980, 1986; Keller et al. 2003). Thus, using the suggested terminology, the yellowthroat can be characterized in the GIS context as a *solid* habitat species where that habitat structure is dense, shrub-level (0–3 m) vegetation with an open overstory. In this way, no special explanations are required when the species occurs in plant communities as disparate but similarly densely vegetated as a regenerating shelterwood cut and a managed switchgrass (*Panicum virgatum*) stand (Murray and Best 2014). Similarly, this approach can be used to explain the occurrence of species associated with various types of edges between landscape components (i.e., edge species), such as the song sparrow's (*Melospiza melodia*) association with deciduous shrubs and saplings adjacent to grassy openings.

We submit that interpreting the concept of patch to include both solid and edge types discernible in remotely sensed imagery (1) represents a more organism-centered, 3-dimensional interpretation of patch as "*a surface area differing from its surroundings in nature and appearance*" and (2) fosters identification of a suite of explanatory variables that meet Kotliar and Wiens' (1990) goal of using objective criteria to define patch boundaries. As such, we believe the concept has great utility, both from an explanatory perspective and from a managerial one, and shall explore its application throughout this discussion.

1.3.5   Guild and Assemblage

Prior to Root's (1967) introduction of the term guild, the common organizational level above individual species was that of a community, discussed below. However, due mostly to their inherent complexity, communities have been referenced and analyzed most frequently in terms of constituent plants or animals, such as forest, marsh, bird, or small mammal communities.

Similarly, despite its original pan-taxonomic definition by Root (1967) as a group of species using a resource in a similar way, guilds also have been analyzed primarily within taxa (e.g., a guild of frugivorous birds). Wiens (1989b) and Morrison and Hall (2002) noted this arbitrary (taxonomic) assignment of species to guilds, as had Jaksic (1981) and MacMahon et al. (1981) earlier. However, Wiens (1989b) also noted that a guild-based approach, despite these potential limitations, could allow development of more predictive models that add a greater degree of formality and defensibility to management decisions (see Appendix B). Thus, although in practice, quantitatively identifying true guilds in nature or objectively assigning members to them a priori is difficult, if not impossible, we agree with Wiens (1989b) that applying the concept, particularly in habitat management, can be useful.

Morrison and Hall (2002: 47) later suggested using the term *assemblage* in place of either guild or community *"when one is simply studying some group of species for any number of interesting reasons."* To date, assemblage has been largely confined to describing animal taxa (e.g., bird assemblages (Bohm and Kalko 2009), small mammal assemblages, insect assemblages, etc.), while the term community continues in use to describe both animal (e.g., coral reef) and vegetation associations (e.g., grassland).

1.3.6 Community

Community, itself, is a term which, although more widely used than almost any other in ecology, has some conceptual and definitional issues, particularly when it comes to defining the boundaries of a community (see Looijen 1998 for a history and discussion). As we noted earlier, this boundary issue also applies to the landscape ecology term "patch," which is typically used as a surrogate for "plant community" in describing discreet mapping units of vegetation on remotely sensed imagery. For our purposes, we will rely on the definition of a community as an interacting set of individuals of multiple species within an essentially uniform set of distinct biotic and/or abiotic conditions (i.e., within the biotope) (Kendeigh 1961; Odum 1971; reviewed by Looijen 1998; Morrison and Hall 2002).

Several other terms in community ecology have related meanings to the term community. The wildlife term "cover type" and the forestry term "stand type" are analogous to the ecological plant community, as is the common use of the term patch in landscape ecology, as noted above. However, cover type and landscape patch also are commonly the equivalent of a biotope [i.e., they may include abiotic components (e.g., a marsh)].

Chapter 2
Image Resolution: Habitat Selection Scale in a Remote Sensing Context

2.1 Scale of Habitat Selection

Researchers have long sought to understand what cues are used by animals to select those subsets of the landscape that maximize their reproductive fitness (MacArthur and MacArthur 1961; Hilden 1965; James 1971; Fretwell and Lucas 1972; Johnson 1980; Cody 1981; Morris 1987; Wiens et al. 1987; Wiens 1989b; Martin 1992; Pribil and Picman 1997; Thompson and McGarigal 2002; Ahlering and Faaborg 2006). Historically, habitat selection studies focused on local structure of the environment and identified selection mechanisms linked to these local (high resolution) conditions. Many studies still employ this ground-based approach. In their review of the influence of variables measured at different scales, Mazerolle and Villard (1999:119) noted, *"One would intuitively expect that patch (local) conditions would be better predictors of species presence or abundance than landscape context (i.e., habitat must be favourable in order for the species to be present)."* Results of the 61 studies they reviewed at that time supported this view. More recently, however, studies employing GIS metrics have expanded in spatial extent well beyond plots to include regional scale attributes (e.g., McGarigal and McComb 1995; Saab 1999; Cushman and McGarigal 2002; Drapeau et al. 2002; MacFaden and Capen 2002; Pearman 2002; Lawlor et al. 2004; Miller et al. 2004; Bakermans and Rodewald 2006; Rioux et al. 2009; Cornell and Donovan 2010; Kennedy et al. 2011; LeBrun et al. 2012). Studies examining behavioral influences on habitat selection such as conspecific attraction (e.g., Campomizzi et al. 2008) also have been added to this mix (reviewed by Ahlering and Faaborg 2006).

A growing number of these more recent studies, while often finding correlations with plot- or local-scale variables (e.g., Lichstein et al. 2002a, MacFaden and Capen 2002), have convincingly argued that consideration of effects at one or several geographically more extensive scales, particularly in anthropogenically

© The Author(s) 2014
J.K. Keller and C.R. Smith, *Improving GIS-based Wildlife-Habitat Analysis*,
SpringerBriefs in Ecology, DOI 10.1007/978-3-319-09608-7_2

influenced landscapes, is important in developing conservation strategies (e.g., Saab 1999; Howell et al. 2000; Johnson et al. 2002; Lee et al. 2002; Thompson et al. 2002; Bakermans and Rodewald 2006; Kennedy et al. 2011; Thompson et al. 2012). Thus, inclusion of GIS metrics in development of species–habitat models has become the norm. This trend notwithstanding, analyses that include spatially more extensive areas have frequently produced models with only modest predictive power (McGarigal and McCombs 1995; Dettmers and Bart 1999; Cushman and McGarigal 2002; Drapeau et al. 2002; Lichstein et al. 2002a; McFaden and Capen 2002; Miller et al. 2004; Thogmartin et al. 2004b, Betts et al. 2006; Howell et al. 2008; LeBrun et al. 2012; Garvey et al. 2013; Lapin et al. 2013). In this and the following chapter, respectively, we discuss (1) issues related to image resolution and (2) the types of metrics employed as potential causes for this outcome.

2.2 Image Resolution and Minimum Mapping Unit Size

First, in describing the attributes of raster-based GIS's, Keller et al. (1979a) and Turner (1989) noted that such programs used cells (minimum size delineation) of a fixed size for the chosen scale of imagery, thus producing variables (metrics) that are relative rather than absolute measures of landscape heterogeneity. Keller et al. (1979a) noted, however, that the value of these metrics is not diminished as long as the resulting cell size, and the underlying image resolution associated with that cell size, has biological relevance to the species or taxon of interest. Although this observation was made originally in describing raster-based programs, all GISs, raster- or vector-based, have minimum size delineation units or minimum mapping units, and associated classification systems (see next section) that are subject to these constraints.

More recently, in analyzing the influence of pixel (or more correctly, *cell*) size variation on landscape pattern metrics, Trani (2002) demonstrated that application of increasingly coarser resolutions (Table 11.2, op. cit.) routinely led to measurement errors of landscape predictor variables (e.g., edge length, interspersion), which reduced species occurrence prediction rates and resulted in misinterpretations of species–habitat relationships (see also Wu 2004; Ostapowicz et al. 2008). She concluded (Trani 2002:151) that *"Only at the proper spatial scale* (sic—resolution and minimum mapping unit size) *do metrics have potential meaning for resource managers."* She went on to note (op. cit.:153) *"Pattern analysis is most useful when the scale of analysis matches the scale at which species use the landscape."* Thus, for any study involving wildlife habitat analysis, the choices of image resolution, minimum mapping unit, and the associated classification system are critical to one's ability to identify and map features/components of potential importance to habitat selection by the organism (Huston 2002; Arponen et al. 2012) and should be based on some understanding of the focal taxon's mechanism(s) of habitat selection and use (see Morrison et al. 1998 for discussion).

Consider, for example, the application of Trani's findings to migratory birds, which are widely agreed to exhibit broad-level selection at geographic scales much larger than an individual territory or home range. Avian visual acuity, however, suggests birds likely accomplish this at much higher levels of *resolution* than the imagery employed in studies designed to test for such assessment (cf. Lawler et al. 2004 in Sect. 2.3).

Migrants associated with shrub-scrub plant communities, for example, have the ability, whether within areas of extensive forest or other unsuitable biotopes, to locate small and ephemeral parcels of such early successional habitat (Confer and Knapp 1979; Titterington et al. 1979; DeGraaf 1991; Probst and Weinrich 1993; King et al. 2001; Keller et al. 2003; Martin et al. 2007; Schlossberg and King 2009). Yet, on 30 m or similarly lower resolution imagery such parcels are typically indiscernible and/or misclassified to more general forest types (Smith et al. 2001; Luoto et al. 2004; Thogmartin et al. 2004a). As a result, as noted by Huston (2002), although a species might be present only within a small subset of a larger sampling area, and thus counted as present within that area, measurement of habitat heterogeneity (i.e., the grain) at a low resolution would fail to differentiate the subset from the average conditions of the whole area, despite the likelihood that actual differences exist. This leads both to inaccurate predictions of species occurrence (Lebbin 2013) and to misunderstanding of the processes that resulted in the observed pattern of occurrence within the larger area (Van Horne 2002). Orians and Wittenberger (1989) summarized this problem, noting that if inappropriate scales of sampling and analysis are used, key factors in species-environment relations may be missed.

While we acknowledge that the resolution of the remotely sensed imagery selected is frequently influenced by (1) the cost of data acquisition or transformation and (2) data availability (McDermid et al. 2009), using an image resolution and associated minimum mapping unit appropriate to address the research question of interest are essential to identify the potentially most insightful predictor variables for exploration of species–habitat or higher level relationships (see also Risser 1987; Wiens 1989a; Wiens and Milne 1989; Mazerolle and Villard 1999; Franklin et al. 2002; Young and Hutto 2002) As noted by Wagner and Fortin (2005), landscape metrics are highly sensitive to scale (sic—resolution and minimum mapping unit), and mapping errors increase if imagery scales are insensitive to the level of proposed classification systems. Ultimately, choice of image resolution and an associated minimum mapping unit are critical to strength, interpretability, and applicability of the results.

2.2.1 What About Mapping Scale?

The researcher's need to visualize and interpret image heterogeneity at an ecological scale thought to be relevant to the organism determines the importance of mapping scale. For example, viewing a remotely sensed image at a mapping scale, as well as a resolution, that allows identification of components reasonably perceived and responded to by the organism can enable both selection and biological interpretation of potentially important predictor variables, points we discuss in Chap. 6. We submit these latter processes are both more difficult if one examines only (1) imagery at an overly small mapping scale or (2) landscape metric output (e.g., land-cover proportions) in isolation, regardless of the image resolution employed. Therefore, although resolution (GRD, NIIRS Level) is far more relevant, the ability of the researcher to observe it (mapping scale) also may be important to consider. Current GIS platforms allow for ready manipulation of mapping scale.

Given adequate image resolution, mapping (image) scale also determines (1) the ability of the researcher to develop a classification system or identify/modify an existing one to characterize landscape components of interest if the image is unclassified and (2) manually or through automated interpretation apply the classification to create the mapped data.

2.3 Landscape Component/Land-cover Classification

Knowing how animals use their habitat is critical to interpreting the results of GIS-based analyses (Van Horne 2002). However, this knowledge is equally critical at the outset of the investigation (cf. Farrell et al. 2013). This is because the values of all GIS-derived metrics will reflect the underlying image resolution and the corresponding level of detail in the chosen classification system. Therefore, both the image resolution and classification system should consider the minimum size landscape components thought to influence habitat choice by the focal taxon. Understanding the focal taxon's life history and habitat use are essential in this process.

For larger or wider ranging species (e.g., bears), landscape component classification may be more general and may be adequately represented by traditional broader cover types such as forest or grassland. For smaller species with typically more limited home ranges or territories and concomitantly finer scales of habitat selection and use, the classification may need to include more discrete elements such as individual shrubs, small canopy openings, or a narrow water feature only a few meters or less across. In either case, image resolution and the accompanying classification system should reflect the focal taxon's response to the subset of the landscape it uses in order to allow biologically meaningful interpretation of the results of GIS-based analyses. If employing a pre-classified image, one should evaluate whether the resolution and classification system of the image are appropriate to meet the objectives of the study (O'Neill et al. 1996; Gallant 2009; McDermid et al. 2009).

Many classification systems and taxonomies have been developed for interpreting and classifying remotely sensed imagery, often on an individual study basis. Among those classifications are a number of widely recognized systems in routine use, often applied to higher resolution imagery, such as aerial photography or Landsat Thematic Mapper, with diverse land-cover or landscape component categories at a range of mapping scales. Some examples include the early land-cover classification suggested by Anderson et al. (1976), the national wetland classifications of Cowardin et al. (1979), ecological land types (ELTs) used primarily by the US Forest Service (Eyre 1980; MacFaden and Capen 2002), the National Land Cover Dataset (NLCD, Loveland et al. 1991), and the National Vegetation Classification Standard (NVCS) developed by the Federal Geographic Data Committee in close collaboration with the Vegetation Classification Panel of the Ecological Society of America (Federal Geographic Data Committee 2008).

In particular, the NVCS embodies a number of desirable attributes important for researchers working with species–habitat relationships to consider. In general, researchers and conservationists would do well to embrace and apply established standards such as the NVCS whenever possible rather than creating their own ad hoc classification schemes. If study-specific classifications of remotely sensed imagery are deemed necessary, they should be developed in close collaboration with remote sensing specialists and vegetation scientists (cf. McDermid et al. 2009).

Lawler et al. (2004) amply demonstrated the strikingly different results that can obtain by employing different land-cover classifications at a particular image resolution, in their case 160 land-cover types (fine grained) versus 14 land-cover types (coarse-grained). They also noted that, for many applications, classification systems should be designed to address species-specific habitat requirements and that "*determining the proper resolution of any classification needs to be incorporated into the design of studies with the same care that recent research has shown must be paid to the effects of scale*" (op. cit.:519). We agree, and note that, as stated in the previous section, the ability to classify cover types or more finely, individual landscape components, is ultimately limited by the *resolution of the image*. In the many GIS-based analyses we reviewed, neither the influence of the classification system chosen nor the limitations on classification accuracy due to image resolution are discussed by more than a few authors (e.g., Bart et al. 1995; Dussault et al. 2001; Smith et al. 2001; Thompson and McGarigal 2002; Hines et al. 2005; Thogmartin et al. 2004a; Davis et al. 2007; Habibzadeh et al. 2013), despite the fact that formal methods for addressing the latter issue exist (Congalton and Green 1998; cf. Bock et al. 2005). Gallant (2009) opined that most databases lack formal accuracy assessments because of the time and cost to complete them (see also McDermid et al. 2009).

As Smith et al. (2001) noted in applying Landsat-5 Thematic Mapper imagery at a nominal resolution of 10 ha for the New York Gap Analysis Project, although overall map accuracy was 74 %, producer accuracies for agricultural land-cover types and shrublands were consistently lower than the average. This is because at this resolution, agricultural cover types are highly dynamic in their spectral responses and shrublands can be confused with other forest types (McDermid et al. 2009; cf. Lapin et al. 2013 inability of Landsat to distinguish cutover from mature spruce). As a result, Smith et al. (2001) cautioned that any interpretations, conclusions, or management recommendations based on habitat associations of (often declining) breeding bird species associated with these early successional cover types (Mitchell et al. 2000; Dettmers 2003) should take into account the mapping accuracies achieved. Thogmartin et al. (2004a) offered a similar caution on the use of the NLCD and McDermid et al. (2009) commented on the quality versus the *utility* for a particular wildlife study of different map sources available to researchers. These studies and the aforementioned lack of explicit consideration of mapping accuracy or other measures of quality in many studies we reviewed, suggest the need for field biologists and conservationists to possess a better understanding of the subtleties, strengths, and weaknesses inherent in the compilation, application, and interpretation of remotely sensed data (Glenn and Ripple 2004; Wulder et al. 2004; Gallant 2009).

2.4 Image Resolution and Interpretability

Avery and Berlin (1985) discussed the interpretability of various mapping scales of remotely sensed imagery and noted that images at a scale of less than 1:10000 precluded identification of individual trees and shrubs, or estimation of their heights. In addition, they and Miller (1996) noted that for satellite imagery (typically 1:40000 at 30 m × 30 m resolution), even classification systems and general mapping of forest cover are problematic due to the lower resolution of this imagery (see also Bart et al. 1995; Dussault et al. 2001; Glenn and Ripple 2004; Arponen et al. 2012).

Wulder et al. (2004) reviewed advancements in data acquisition technology (greater spatial, radiometric, temporal, and spectral resolution) for their utility in ecological studies. While they acknowledged the improved interpretive capabilities of modern sensors, they noted that only when there are many pixels per object rather than many objects in a single pixel can the object of interest (e.g., individual trees, and shrubs = grain size for our discussion) be meaningfully characterized (but see Bock et al. 2005). This H-resolution imagery (Strahler et al. 1986) allows identification of a high degree of local heterogeneity (i.e., multiple landscape components) and contains large amounts of spatial information (McDermid et al. 2010). In contrast, L-resolution images (e.g., a Landsat image that contains multiple trees within a single 30 m pixel) contain less spatial information about individual trees or other landscape components of potential interest but would provide an acceptable level of information about forest distribution at the scale of an entire stand (op. cit.). These are critical points in the investigation of species–habitat relationships using GIS, yet appear to be rarely considered explicitly.

The limitations of decreased image interpretability and increased misclassification at lower resolutions (Franklin et al. 2000; Fleming et al. 2004; Thogmartin et al. 2004a; Hines et al. 2005) are magnified when one considers the fundamental differences in the scales at which wildlife species select habitat (Wiens and Milne 1989; Pearson and Gardner 1997). Kerr and Ostrovsky (2003) observed there often is a perceived mismatch between the data sought by ecologists and the data collected with remote sensing instruments. Turner et al. (2003) suggested this perception is declining due to the increasing availability of high-resolution data that can be directly linked to traditional field-based ecological measurements. Although we agree that imagery with appropriate resolution (measured as GRD or NIIRS Level) may be available, the resolution of imagery (or mmu) actually chosen for use in a particular analysis does not always account for differences in the scales at which wildlife select habitat (O'Neill et al. 1986; Gottschalk et al. 2005; cf. Donovan et al.'s (2012) use of the same L-resolution imagery for Bobcat [*Lynx rufus*] and Ovenbird [*Seiurus aurocapillus*]).

In contrast to this latter study, Hutto (2014) used H-resolution aerial photography (Fig. 1, 1:1560 scale, no resolution given) to identify "every visible shrub, tree, and downed log" (op. cit.:123) while mapping individual perch sites and territory boundaries of male Calliope Hummingbirds (*Selasphoris calliope*) on a dispersed lek within a 20-year-old seed tree cut in western Montana. Similarly, Farrell et al. (2013) combined LiDAR (resolution to less than 1 m) with H-resolution (0.35 m) color-infrared scanned image data (SID) to develop predictive occurrence models for two listed species of passerines. Conversely, using the same dependent data set, models they developed that included only traditional canopy cover estimates from L-resolution imagery were not plausible for either species.

Lawler et al. (2004) further illustrated this resolution/interpretability relationship. Using Breeding Bird Survey (BBS) data for six different species of birds, they compared models of species occurrence within the conterminous US built with fine-grained versus coarse-grained image classification data based on satellite imagery with 1.1-km^2 resolution. Most (8 of 12) models, whether employing the coarse- or the fine-grained cover type classification, included both cover type and climatological variables to explain the species distributions. However, none of the remaining four models, (2 each) for the House Wren (*Troglodytes aedon*) or Savannah Sparrow (*Passerculus sandwichensis*), included any of the available land-cover types or 24 pattern metrics as explanatory variables, only climate data.

The authors offered that the lack of inclusion of any land-cover variables in the House Wren or Savannah Sparrow models suggested a failure of the land-cover classification system to capture the habitat preferences [sic] of these species. They suggested (op. cit.:528) that "...*accurate models for some individual species will require the definition of species-specific land-cover classifications designed to address specific habitat requirements*". We agree (see our sample analysis in Chap. 6 and associated landscape component classification in Appendix C). However, we also suggest that use of the landscape by these species was simply at too fine a resolution to be detected using the 1.1 km^2-resolution satellite imagery employed (e.g., note the reference in the foregoing quote to "species specific *land cover classes*", which connotes the mismatch of the imagery resolution employed to the scale at which small passerines actually use the landscape). Thus, although this type of range-level analysis can be useful in interpreting distributional patterns over large geographic scales (e.g., Collier et al. 2012; LeBrun et al. 2012; Lapin et al. 2013), it also may produce misleading or incomplete model results, as above; and appears generally inappropriate for identification of local-scale management or conservation approaches due to the mismatch of image resolution and associated minimum mapping unit size with species' habitat-use scale, classification system notwithstanding (Farrell et al. 2013). We further discuss the use of BBS data and GIS sampling of BBS routes in Chap. 4.

Analytical scale disconnects as described above are not exclusive to vertebrate studies. For example, Rykken et al. (1997) attempted to correlate ground beetle distributions to the US Forest Service's Ecological Land Types (see McFaden and Capen 2002), whose minimum mapping unit is 10–100's of ha in size. Not surprisingly, based on the preceding discussion, no relationship was found (but see Luoto et al.'s 2002 use of satellite imagery to successfully model habitat for a more wide-ranging insect).

2.5 Matching Image Resolution to the Scale of Species Function

2.5.1 Domains of Scale

In a discussion of scales of analysis, Wiens (1989a) noted that patterns evident at a biogeographic scale reflect underlying patterns and processes at a local scale; and therefore, an understanding of local-scale events is necessary to properly interpret coarser scale patterns. Huston (2002:10) restated this observation in terms of resolution, describing how "...*patterns can be detected at resolutions far coarser than the resolution needed to understand the processes that produce the pattern, which is also the resolution needed to predict the pattern*." So, what is the "appropriate" scale and (image) resolution of analysis to correctly infer process from observed pattern?

Cautioning against unrecognized bias in such choices, Wiens (1989a:391) then noted that "*scales chosen for analysis by researchers typically reflect hierarchies of spatial scales that are based on our own perceptions of nature. Just because particular scales seem 'right' to us, is no assurance that they are appropriate*" for the particular taxon of interest. Selection of image resolution in GIS-based analyses warrants a similar caution. This is because, as inferred by Huston's (2002) comment above, as image resolution decreases, there is a concurrent decrease in the ability to identify, classify, and quantify accurately species-specific landscape components (grain) and their spatial arrangement (habitat size, configuration, and distribution) (see Smith et al. 2002, 2003; Fleming et al. 2004). This leads to an increasing disconnect between actual occurrence of habitat on the landscape and one's ability to measure it, at any hierarchical scale.

Wiens (1989a, and Fig. 4 therein) then considered scale-dependency in ecological systems theoretically and argued that within the spectrum of potential analytical scales, there are "domains" (i.e., subsets of the scale continuum) of particular ecological phenomena within which process–pattern relationships are consistent, regardless of the scales of observation within that domain. He went on to note that proper analysis requires that the scale of researcher measurements and that of the organism's responses fall within the *same* domain. We submit that viewed in this way, "domain" represents the range of image resolution and level of image classification at which GIS-based habitat analysis should occur. This is because in GIS, the ability to interpret an organism's responses to its environment (i.e., habitat selection and/or use) in a biologically meaningful way depends on adequately quantifying the composition, structure, and spatial arrangement of relevant elements composing that environment, which ultimately depends on image resolution and the associated classification system employed in the analysis.

Given this relationship, a good starting point for determination of appropriate image resolution would seem to be within territory core areas of activities such as foraging, singing, and nesting (e.g., Balbontin 2005; Barg et al. 2005, 2006; Moore et al. 2010; Hutto 2014). That is, one should ensure that the imagery, whether detected passively (e.g., photographs) or by propagated signals (e.g., LiDAR), provides the capability to characterize thoroughly **intraterritory spatial heterogeneity** (cf. Holland et al. 2004). As but one example, St-Louis et al. (2006), using a range of sampling window sizes (i.e., nested, increasingly larger, fixed-sized, typically square samples on a GIS map), found the strongest correlations of bird species richness with landscape heterogeneity (measured as image texture on high-resolution (1 m) aerial photographs) at sampling window sizes (21 × 21 m–31 × 31 m = **441–961 pixels**) that represented subsets of territories for virtually all of the species being studied. Landsat (30 × 30 m = **1 pixel**) or other lower resolution imagery would not afford the same ability to quantify intraterritory heterogeneity. The results of this assemblage-level study illustrate the desirability of applying appropriately resolved imagery not only to individual species but also to higher organizational levels, as well.

With regard to broader geographic analyses, our literature review revealed that although understanding range-wide patterns of habitat occupancy is a desirable objective, attempts to elucidate such patterns for smaller animals using low resolution imagery have met with only limited success (but see Collier et al. 2012). Thus, although it can be argued that high-resolution imagery is not essential when examining context or for multi-scale studies of broad geographic areas, use of lower resolution imagery for smaller species can result in misinterpretations in both types of studies, and ultimately, in misinformed management recommendations.

The foregoing discussion suggests that although habitat selection may proceed in a top-down manor as described by Johnson (1980) and questions asked by researchers at biogeographic scales (e.g., population viability) may differ from those asked at local scales (e.g., presence / absence), the variables and image resolution used to address them should overlap broadly. As noted by Wiens (1989a:390), *"If we study a system at an inappropriate scale, we may not detect its actual dynamics and patterns but may instead identify patterns that are artifacts of scale. Because we are clever at devising explanations of what we see, we may think we understand the system when we have not even observed it correctly."* This astute observation points to the need for equally careful selection of the image resolution and classification system in GIS-based habitat analyses.

Only when image resolution (domain of scale) is appropriate for the species at all levels of analysis (local to biogeographic), can habitat occurrence, its spatial distribution, and hence, associated species distributions and populations, be accurately estimated using GIS-based models (cf. Farrell et al. 2013). Given the computational requirements and cost of using high-resolution imagery or LiDAR over biogeographic scales, sampling locations should be carefully considered when developing species–habitat models with the intent of estimating population distributions across range-wide levels of analysis (cf. Huston 2002; MacKenzie and Royle 2005).

Although it is beyond the scope of this paper to fully examine these issues, they do suggest potentially useful future research (e.g., are there differences in the apparent influence of the matrix when it is viewed at high versus low resolution, using classification schemes appropriate for both?).

2.5.2 Less Arbitrary Selection of Image Resolution

Wiens (1989a) noted that scales chosen for analysis of species–habitat associations are generally arbitrary but that what constitutes an "appropriate" scale is somewhat dependent on the questions asked by the researcher. He then noted that differences among organisms also affect the scale of investigation and suggested that these differences somewhat paralleled differences in body size, which raised the possibility of using allometric relationships to scale analyses. We agree and propose the use of a simple body-size-based rule to aid selection of a less arbitrary, more species-scaled image resolution for any GIS-based habitat analysis.

Considering the foregoing discussion and references cited, we suggest that, as a general rule, the remotely sensed imagery employed to elucidate species–habitat relationships should be capable of resolving landscape elements ≤5x–10x the body length of the species of interest. For example, applying this criterion to a 13-cm warbler would result in a GRD range of 0.65–1.3 m, requiring imagery that could resolve small shrubs or saplings in the range of 1 m in width, a reasonable goal for a small, foliage-gleaning insectivore. However, as with all such generalities, there are exceptions. For example, due to their flexibility, snakes can take cover under landscape elements (e.g., rocks) that are considerably less than their body length, which again underscores the importance of understanding the natural history of the focal taxon. In this case, the smallest resolvable components of interest are likely *less than half* the length of the individual.

In general, we suggest researchers ask the following:

- Will the chosen remotely sensed imagery resolve **landscape components and potential intraterritory heterogeneity** that reflects the focal taxon's scale of habitat use or association that is of interest to the reseacher?
- If the scale of habitat use or association is unclear, are the smallest resolvable landscape components in the imagery more than 5x–10x the body length of the species of interest? Are there morphological or physiological attributes of the species that suggest the need for resolution of even smaller landscape components?

- Regardless of whether the previous questions are answered affirmatively or negatively, what information might be lost at the chosen image resolution?
- Does the classification system employed reflect the component composition and heterogeneity described above?
- Will the chosen resolution allow meaningful characterization of **landscape attributes (i.e., spatial arrangement) of those components?**

Combining the body size–image resolution rule, we offer with an understanding of the species' natural history and core area, territory, or home range size, should suggest the appropriate image resolution and level of land-cover classification (e.g., physiognomic vs. floristic) to apply.

2.5.3 Edges and Ecotones: Resolution Effects on Interpretation of Habitat Association

As implied in the Working Definitions Chapter, the concept of edge has long interested ecologists and edge quantification has been facilitated by the advent of GIS. However, considering edges only at the scale of plant communities (i.e., *ecotones*), which is the typical case, implies that (1) there are relatively few types of edge in a landscape and (2) individually, those edges are relatively extensive spatially (cf. Fig. 1.1). As noted, the alternate view considers that myriad edges of numerous types are present at many scales (e.g., Fig. 1.2), both between and frequently within plant communities (Risser 1987; Wiens 1989b). Under this latter concept, an ecotone such as that between a forest and a hayfield represents but one general type of interface or boundary, which itself contains multiple edge types (Gosz 1991) recognized by various species or guilds of wildlife. Along this extensive-to-local continuum of edge scale, increasingly smaller animals use progressively more fine-grained edge types (see Kotliar and Wiens 1990).

For example, White-tailed Deer (*Odocoileus virginianus*) and Red-tailed Hawk (*Buteo jamaicensis*), both of which have home ranges on the order of several hundred to >1,000 ha (e.g., Tierson et al. 1985; Andersen and Rongstad 1989), might be expected to select habitat generally at the scale of plant communities and ecotones [but see suggested finer resolution selection for deer by Fleming et al. (2004)], whereas passerines (breeding territory = 0.5–2 ha) appear to select at the scale of individual landscape components such as trees, shrubs, openings, and edges between them (cf. Suarez et al. 1997; Barg et al. 2006). Smaller or more sedentary animals (e.g., small mammals, salamanders, small fish, mollusks, gastropods, spiders, insects = home ranges of <0.5 ha) have been demonstrated to select at even finer scales of landscape components and associated edges (e.g., rocks, logs, individual plant parts; cf. Krawchuk and Taylor 2003; Stoddard and Hayes 2005; McKenny et al. 2006; Ewers et al. 2007; Matias et al. 2007; Macreadie et al. 2010; Matias et al. 2010; Moore et al. 2010; Vierling et al. 2011).

Failure to consider the resolution at which habitat selection occurs when attempting to categorize species as "edge," "forest interior," or other habitat association, has led to misinterpretation of these associations, and can have profound implications for conservation and management (Franklin et al. 2002). For example, the Cerulean Warbler (*Setophaga cerulea*) often has been characterized as a "forest interior" species. Yet, recent evidence suggests that it is frequently associated with closely spaced, variously sized canopy gaps (i.e., edge) (Weakland and Wood 2005; Perkins 2006; Bakermans and Rodewald 2009; McElhone et al. 2011; Boves et al. 2013; Perkins and Wood 2014) caused by tree deaths (more frequent in floodplains), blowdowns (more frequent on steep slopes and ridge tops), or deliberate forestry practices such as shelterwood harvest (Carpenter et al. 2011; S. Stoleson, USDA Forest Service personal communication). Thus, when viewed at the within-territory scale, it may actually be more of a forest interior-edge species, suggesting previously unconsidered management approaches (cf. Perkins and Wood 2014).

Imbeau et al. (2003) touched on the "labeling" issue, noting that categorizations of some bird species as early successional in one research context conflicted with classifications of the same species in another context, such as when defining so-called edge (=ecotone) species (see also Miller et al. 2004). They argued that early successional species are actually shrubland specialists and occur at mature forest edges (ecotones) only due to lack of appropriate shrubland habitat elsewhere in the immediate vicinity. They further argued (op. cit.:514) that "to be considered a true edge species, a species has to require the simultaneous availability of more than one habitat type" (i.e., habitat *sensu* plant community). As a result, they concluded that "real edge species" (i.e., ecotonal edge) are probably quite rare. We argue instead that edge species are only rare if the definition of edge is confined to the spatial scale of an ecotone. When considered at a species-specific scale and resolution (i.e., grain), the number of species dependent on edges between different habitat components (e.g., trees, shrubs, grass, and water) is, in fact, quite high. Using our Working Definition of edge habitat (Chap. 1), Red-tailed Hawk, Song Sparrow, Golden-winged Warbler (*Vermivora chrysoptera*), and Praying Mantis (*Stagmomantis carolina*) are all edge species, each associated with edges at spatial scales commensurate with their size and habits.

Additionally, some researchers have suggested that amount of edge is more influential on bird communities at the landscape scale (McGarigal and McComb 1995; Hagan et al. 1997). We suggest that when edge is measured at the plot (i.e., habitat) or biotope/community scale, it can be at least as predictive of community (assemblage) composition and potential ecotonal effects via correlations of more specific edge types with individual species or guilds. This has rarely been done to date (but see Keller 1990; Keller et al. 2003; Chapa Vargas and Robinson 2007; Rehm and Baldassarre 2006; Macreadie et al. 2010; Chap. 6 herein).

For example, as part of a study of the use of even-aged seres by breeding birds following cutting, Keller et al. (2003) analyzed edge effects using 1:5000 stereoscopic aerial photography at a mapping scale of 1:2000 with resolution of <0.75 m (NIIRS Level 6). They found that in addition to the perimeter edge (i.e., ecotone) between cutover areas and adjacent forests, which would be generally identifiable in Landsat imagery, clear-cuts initially (i.e., post-cut years 1–4) contained high levels of internal edge (e.g., shrub-grass and sapling-grass). At an ecological scale relevant to habitat selection by passerines, the density of these latter edges was strongly associated with early successional species of terrestrial gleaners, primarily sparrows. As succession proceeded, regenerating sprouts and root suckers filled in the openings, thus eliminating the internal edges, and the sparrows were replaced by foliage-gleaning insectivores, primarily warblers. Viewed at a lower resolution Landsat scale, these internal edges would have been undetected. Only the clear-cut itself and surrounding forest would have been identified, and thus, only the plot perimeter edge (ecotone), which remained unchanged, would have been quantifiable (cf. Schlossberg and King 2008), setting up spurious correlations with edge-associated guilds such as the sparrows, which occurred primarily *within* the clear-cuts, not along the perimeters.

The preceding series of examples illustrates how edge-associated species may cue on ecotone-scale edges (e.g., Red-tailed Hawk and White-tailed Deer), intermediately scaled edges (e.g., flycatchers = canopy-opening or forest gap; song sparrow = shrub-grass) or much more localized subsets of edges [insects = edge between a leaf and adjacent air (Krawchuk and Taylor 2003) or small fish = edge between seagrass and adjacent open water (Macreadie et al. 2010)]. At the low resolution, large geographic scales typically employed in GIS analysis of species–habitat relationships (e.g., Landsat = 1:40000 with GSD 30 m, or widely available aerial photography databases = 1:24000), even intermediate scales of edge are generally not discernible and are thus, unmeasurable (Avery and Berlin 1985; Wulder et al. 2004).

Of the species mentioned above, lower resolution imagery (GRD <30 m, NIIRS Level 3 or higher) would appear reasonable for development and application of habitat models to only the larger-bodied, wider ranging species such as the deer and hawk (cf. Laymon and Reid 1986; Nixon et al. 1988; Poppelwell et al. 2003; Nielsen, et al. 2006; Davis et al. 2007; Kays et al. 2008; Rioux et al. 2009; but see Palmeirim 1985; Huber and Casler 1990; Thompson and McGarigal 2002; and Fleming et al. 2004; reviewed by Gottschalk et al. 2005). Higher resolution (e.g., GRD <1.2 m, NIIRS Level 5 or higher) imagery, and an associated larger mapping scale (e.g., 1:5000) to facilitate researcher interpretation, would be more appropriate for passerines (see our analysis in Chap. 6), and very high-resolution imagery (e.g., GRD <0.4 m, NIIRS Level 7 or higher) generally would seem necessary for small mammals, salamanders, and most arthropods, at least those associated with open-canopied habitats (e.g., Cronin 2009; see also scale of LiDAR imagery for spiders in Vierling et al. 2011).

As suggested by this discussion, failure to adequately match image resolution to the resolution at which species perceive and respond to edges (or solid patches) in the landscape can result in derived metric values that are artifacts of inappropriate image resolution (Wiens 1989a). Furthermore, even with higher resolution imagery that may more closely match landscape component composition and resolution with organism use of the landscape, GIS metrics (or researchers) frequently equate (i.e., lump together) all edge types or reduce specific edge types to "levels of contrast," with a concomitant a priori loss of information (see Chap. 3).

Collectively, inappropriate resolution and overly simplified interpretations of edges can produce completely misleading correlations with metrics such as fractal dimension or edge density, which can result in incorrect inferences about species–habitat associations. Ultimately, these shortcomings can have important consequences for management and conservation.

2.6 Landsat Versus Higher Resolution Imagery

Many biotopes, despite their definition as being "more or less uniform," are matrices of components containing substantial internal heterogeneity, quantification of which is entirely resolution dependent since an apparently solid habitat type, with problematic or unmeasurable edge at one resolution, reveals that edge (i.e., internal heterogeneity) at a higher resolution. In ecological terminology, this is the change in *grain* size, the smallest element used in habitat selection by the organism. This unresolved heterogeneity is very typical of Landsat imagery when considering open-canopy plant communities such as shrub-steppe, oldfields, early stage clear-cuts or shelterwood cuts, and some grasslands (cf. Bellis et al. 2008).

Figure 1.1, an image of central New York, illustrates the type of broad scale heterogeneity that has been quantified in many avian studies employing Landsat or other remote sensing data with similar resolution (e.g., Donovan and Flather 2002; MacFaden and Capen 2002; Betts et al. 2003; Thogmartin et al. 2004b; Howell, et al. 2008; Cornell and Donovan 2010; LeBrun et al. 2012). Such imagery and the resulting GIS data sets characterize many passerine territories with only a few pixels, each of which, with a sample area frequently of 900 m^2, may actually include diverse habitat components ("mixed pixels") now reduced to a single, sometimes misclassified, cover type (Fig. 2.1, also see Sect. 2.4 and Appendix B) (Bart et al. 1995; Congalton and Green 1998).

As shown in Fig. 2.1, even a modest 10-m GSD mapping resolution using high-resolution (NIIRS Level 6) aerial photography produces 9x more information (81 cells vs. 9) than the 30-m GSD mapping resolution provided by Landsat for a nominal 0.81 ha passerine territory. Additionally, this resolution allows much more accurate classification of the landscape components actually present on the ground. Equally importantly, the 10-m versus 30-m cell size allows more accurate characterization of the spatial arrangement of now identifiable, and thus classifiable, individual landscape components such as trees, shrubs, and grass in an oldfield or clear-cut (see Chap. 3).

The arrangement of these components dictates the presence or absence of declining early successional species such as Golden-winged Warbler (GWWA) and would be unquantifiable at a lower image resolution such as Landsat even if the oldfield or clear-cut itself were identifiable (Fig. 1.1). This is also true for the increasing number of studies employing widely available 1:24000 aerial photography. In the case of the GWWA and other Parulidae, some researchers have suggested their habitat selection cues are at even finer scales, on the order of 10 m^2 or

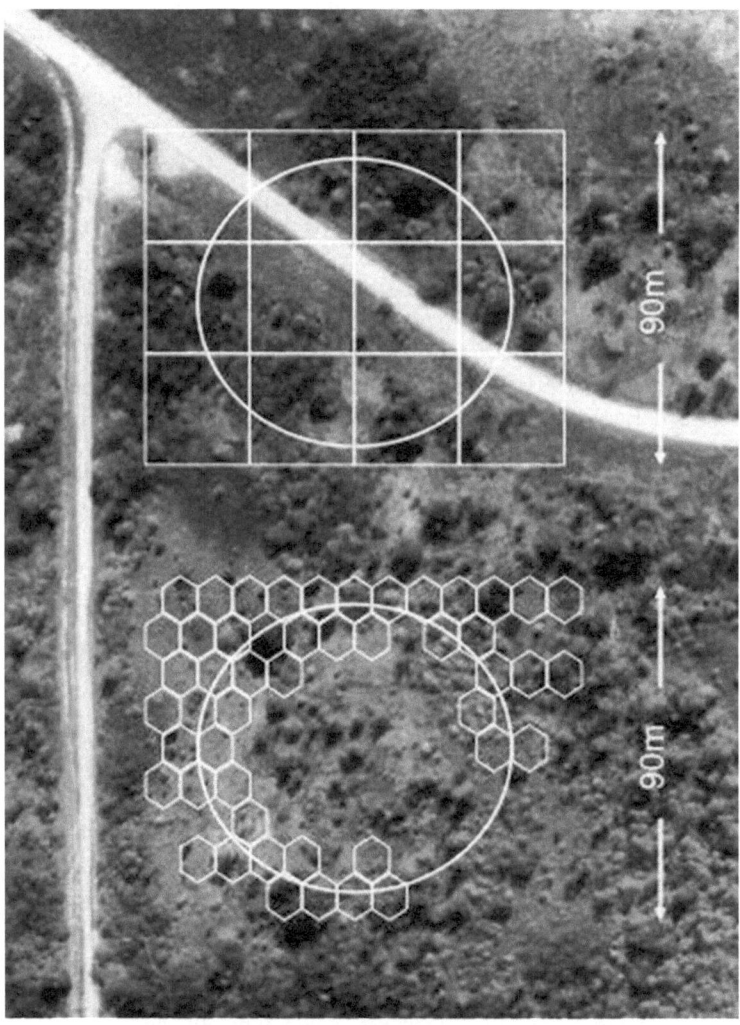

Fig. 2.1 Area of detail in Fig. 1.1 with grids of two cell sizes, 100 m² hexagons and 900 m² squares, superimposed over nominal 0.81 ha passerine territories (ovals) on 1:2000 aerial photography with <0.75 m (NIIRS Level 6) resolution. Notice in comparison with Fig. 1.1 how this much higher resolution image viewed at a larger scale allows identification of landscape components such as individual trees, shrubs, and small patches of open grass thought to be used in habitat selection by passerines. Note also the increased classification difficulties at the greater 30-m ground sampling distance (GSD = cell to cell center distance) of the Landsat-scale cells on high-resolution imagery due to increased inclusion of multiple identifiable component types (e.g., deciduous sawtimber, deciduous saplings, deciduous shrubs, coniferous trees, bare ground, and herbaceous cover) within a single 30 × 30-m (Landsat pixel equivalent) cell. *Source 22* May 1977 aerial photography from an altitude of 1,100 m using a Hasselblad camera and 70-mm black and white film

less (Jeff Larkin, IUP, personal communication; Carpenter et al. 2011). Thus, mismatches of organism space use to image resolution will inevitably lead to lower correlations of species with landscape metrics intended to explain their occurrence (Trani 2002). This is exemplified in many of the studies we reviewed where Johnson's (1980) Third Order (habitat) analysis is attempted using Second Order (biotope) resolution imagery. This suggests the advisability of conducting exploratory analyses (James and McCulloch 2002) employing the highest resolution imagery available (compare Figs. 1.1 and 2.1). Lastly, as noted earlier, alternatives to the use of higher resolution imagery include textural analysis and LiDAR, both of which allow characterization of the heterogeneity within a given landuse class (e.g., Bellis et al. 2008; Graf et al. 2009; Seavy et al. 2009; Goetz et al. 2010; Vierling et al. 2011; Farrell et al. 2013).

Ultimately, interpretation of a species' habitat or patch association, or an assemblage's plant community as "solid" or "edge" is dependent on (1) the resolution at which the landscape is viewed by the researcher (i.e., resolution and mapping scale of imagery) and (2) the resolution-dependent explanatory variables used to test the association. Both should always attempt to match the focal taxon's use of that landscape (Keller et al. 1979a; Keller 1986; O'Neill et al. 1986; Turner 1989; Wiens 1989b; Noss 1991; Orians and Wittenberger 1991; Morrison et al. 1998; Dettmers and Bart 1999; Mazerolle and Villard 1999; Potvin et al. 2001; MacFaden and Capen 2002; Thompson and McGarigal 2002; Trani 2002; Barg et al. 2005; McGarigal and Cushman 2005; Moore et al. 2010; LeBrun et al. 2012; Farrell et al. 2013).

2.7 Gaps in Information: Scale Disconnects Between Local and Landscape-scale Metrics

Few studies measure landscape variables at a resolution that would allow quantification of intraterritory heterogeneity or, in the case of open-canopy communities, even within-biotope heterogeneity (but cf. Thompson and McGarigal 2002; Goetz et al. 2010; Farrell et al. 2013). Researchers frequently either (1) choose a minimum landscape component size (MMU), often a plant community, equivalent to the size of an *entire territory* used by the focal species (e.g., Magarigal and McComb 1995; Cushman and McGarigal 2003) regardless of whether the image resolution might allow a finer level of classification, or (2) use lower resolution imagery that limits the minimum resolvable landscape component to something more spatially extensive, again typically an entire plant community, than the scale of individual components (e.g., trees, shrubs, grassy openings = grain) used by the organism to select habitat (e.g., Rykken et al. 1997; Lee et al. 2002; Lichstein et al. 2002; Betts et al. 2003, 2006, 2007). In either case, the choice (mmu and/or image resolution) completely precludes GIS measurement of intraterritory heterogeneity or, at a minimum, restricts its characterization.

In addition, when hierarchical studies include on-the-ground measurements of habitat, the jump to a spatially more extensive and coarser-grained remotely sensed image may result in a GIS minimum mapping unit that is neither overlapping nor even contiguous with the scale of variables being measured on the ground (but see Saab 1999). This leads to a potential gap in information on the influence of local-scale spatial heterogeneity (cf. With 1994; Hagan and Meehan 2002; Lee et al. 2002; Betts et al. 2006; Dickson et al. 2009; Cornell and Donovan 2010), which, in turn, can lead to misinterpretation of the importance of landscape- versus local-scale variables (see discussion of scale disconnects in Kotliar and Wiens 1990).

In the case of passerines, other small vertebrates and most invertebrates, even most aerial photos are at resolutions too low (NIIRS Level 3) to allow accurate interpretation and thus, quantification, of the individual landscape components important to these species. As a result, if the scale of heterogeneity is not measured at the scale of the organism's response, it can be deemed unimportant to occurrence, even when it really is (With 1994; Mazerolle and Villard 1999; Huston 2002; Trani 2002).

2.8 Multi-scale Analyses

There has been much discussion about the need for examining species–habitat relationships at multiple scales (Pickett and White 1985; Urban et al. 1987; Wiens 1989a; Kotliar and Wiens 1990; Allen and Hoekstra 1992; Virkkala 1991; Forman 1995; Saab 1999; MacFaden and Capen 2002; Thompson and McGarigal 2002; Thompson et al. 2002; Van Horne 2002; Johnson et al. 2004); but what criteria determine the limits of the scales to be considered? Kotliar and Wiens (1990) suggested grain and extent as the lower and upper limits, respectively, of investigation but acknowledged the potential difficulty of identifying these limits. As noted earlier in this chapter, Wiens (1989a) suggested that they could be approximated and we suggested species-specific body-size criteria for establishing grain.

Among many authors commenting on scale effects and the need for analysis at multiple scales, Brennan and Schnell (2007:631) noted that analyses of species–habitat relationships at multiple scales *"allow the data to indicate the most appropriate scale or scales for a particular study, rather than depending entirely on a researcher's subjective perception of what scales are important to a given species."* We agree; however, this also implicitly assumes that both the resolution and range of scales (sampling areas) examined include the most appropriate ones to address the research question for the taxon of interest (see O'Neill et al. 1991; Wiens 2002). We are not sure this always is the case with GIS analyses, certainly for passerines and smaller animals, and refer the reader to Wiens' (1989a) quote in Sect. 2.5.1 regarding identifying patterns that are simply artifacts of scale, or per this discussion, artifacts of image resolution. We submit that frequently, low resolution of the imagery employed limits both identification of meaningful landscape components and quantification of their spatial arrangement (heterogeneity) for the species or assemblage being studied (e.g., Donovan and Flather 2002; MacFaden and Capen 2002; Betts et al. 2003; Thogmartin et al. 2004b; Brennan and Schnell 2007; Howell et al. 2008; for discussions see Trani 2002 and Meyer 2007).

For all of the preceding studies, and many others we reviewed, one can legitimately ask whether the use of (1) higher resolution imagery, (2) smaller minimum size delineations, (3) a more detailed landscape component classification system, and (4) more explicit (i.e., taxon-specific) edge types (cf. Keller 1990 and Chap. 6 herein) might produce better correlations and more biologically interpretable results at both the local and regional scales? In general, there appears to be a field-wide lack of recognition of the potential effects of these issues on resulting management and conservation efforts.

To further emphasize the importance of the initial choices of image resolution, classification system and minimum size delineation, consider the following study of a much larger species than the passerines we have been discussing. Thompson and McGarigal 2002 examined habitat use for Bald Eagle (*Haliaetus leucocephalus*), a species with a home range of well over 1 km^2 (op. cit.), using aerial photography at the relatively large mapping scale (for a bird of prey) of 1:7500. Image resolution was not mentioned, although the minimum mapping unit was 0.01 ha, implying a fairly high-resolution image. Although close to the scale we recommended as a minimum for passerines (1:5000 mapping, NIIRS 5 or higher) in Sect. 2.5, the authors' own analyses still demonstrated a loss of explanatory power for several aspects of habitat use when information on the photos was aggregated to larger minimum size delineations, in what amounted to use of lower resolution imagery. Given the eagle's large home range and, by inference, coarse-grained use of the landscape, this result emphasizes the need to start with the highest resolution imagery available to better assess local-scale relationships when examining GIS-based habitat associations (see also Fleming et al. 2004).

Although landscape-scale (i.e., large geographic extent) influences are clearly real, particularly for less mobile species, we suggest that more careful matching of image resolution and GIS analytical scales to the higher order habitat selection scales (*sensu* Johnson 1980) of the organisms of interest (Figs. 1.2 and 2.1) will lead to explanation of a greater amount of variance in species–habitat associations at local (habitat) and biotope (plant community) scales of analysis (cf. Farrell et al. 2013), particularly in less anthropogenically influenced landscapes (cf. Dickson et al.'s 2009 and LeBrun et al.'s 2012 cautions regarding application of regionally derived models (developed at low resolutions) to local areas).

2.9 Conclusions

- Species–habitat correlations improve significantly as the resolution of the imagery more closely matches the ecological scale at which the organism uses the landscape (Keller 1986; Huston 2002; Trani 2002; Chap. 6 herein). Therefore, use remotely sensed imagery that resolves, at a minimum, the smallest landscape component or component combination (edge) thought to be used in habitat selection by the focal taxon. If the scale of habitat use is unclear, use imagery that can resolve landscape components as small as 5x–10x the body length of the species of interest, unless species morphology or physiology suggests even smaller objects are potentially important.

- We suggest the utility of at least exploring this body size–image resolution relationship, even for biogeographic scale questions of population viability. This is because the ability to identify, classify, and quantify accurately species-specific landscape components, component combinations, and spatial arrangements of these elements, which are often critical to species occurrence, decreases as imagery resolution decreases, resulting in an increasing disconnect between the actual and measurable amount of habitat on the landscape. Use of low resolution imagery, even for questions at large geographical scales, increases the risk of obtaining misleading species–habitat correlations and misinterpreting those relationships.

- When image resolution is adequate, but the classification system is overly general (i.e., similar but identifiable and potentially meaningful landscape components are lumped together within the classification), even geospatial variables may not capture threshold spatial distributions of habitat on the landscape.

- Use the highest resolution imagery available/affordable. One can always aggregate information (cf. Thompson and McGarigal 2002). One can never go back later and measure smaller components on resolution-limited imagery (cf. McElhone et al. 2011; Arponen et al. 2012).

- Landsat data (30 m resolution, square pixels) and smaller scale aerial photography are widely available in digital formats, and analysis software is optimized to deal with them (e.g., ARCINFO, GUIDOS).

- Based on the strength (i.e., % variance explained and predictive capability) of reviewed species-habitat models, Landsat may be adequate for large-bodied or wide-ranging mammals and many raptors; but unless landscapes are "simple" (i.e., relatively homogeneous) and/or the imagery employed has high classification accuracy, it appears generally inadequate for most passerines, small mammals, most herptiles, fish of lower order streams, and many invertebrates.

- The tradeoff in selecting higher resolution imagery and a more fine-grained classification system is one of increased cost and perhaps time versus the loss of information at lower resolutions and coarser classifications.

Chapter 3
Explanatory Variables

What should the explanatory variable measure? In their discussion of scales of patchiness, Kotliar and Wiens (1990) defined grain as the smallest scale at which an organism responds to patch structure, suggesting that in GIS, identification of individual landscape components used as cues by organisms should be the starting point in habitat quantification since, at scales smaller than grain, the environment is perceived as functionally homogeneous. In line with this thinking, which reinforces our discussion of the importance of image resolution, innumerable studies have found individual species to be most frequently associated with specific structural subsets of the biotopes they inhabit (e.g., MacArthur et al. 1962; Pianka 1967; James 1971; Brown 1973; Gorman and Karr 1978; Holmes et al. 1979; Rice et al. 1984).

More recently, GIS-based studies employing landscape metrics have emphasized the spatial attributes of these biotopes, sometimes including abiotic attributes such as slope gradient and aspect, along with the additional influence of the composition of the surrounding landscape (e.g., Saab 1999; Dettmers and Bart 1999; Howell et al. 2000; Cushman and McGarigal 2002; Hagen and Meehan 2002; Lichstein et al. 2002a; Kays et al. 2008; Wilson and Watts 2008; reviewed by Mazerolle and Villard 1999). Regardless of the spatial extent of the analysis, habitat selection has been interpreted as a species response to compositional, structural, and other environmental cues that facilitate its functional role so as to maximize reproductive fitness (Hilden 1965; Holmes et al. 1979; Cody 1981; Wiens 1985, Fig. 5; Martin 1987, 1992; Wiens et al. 1987; Pribil and Picman 1997; Johnson 2007), all of which points to the need to consider carefully the species natural history (Dettmers and Bart 1999; O'Conner 2002; LeBrun et al. 2012) and the scale at which it responds to its environment (Whittaker 1975; Keller et al. 1979a; Addicott et al. 1987; Morris 1987; Wiens 1989a; Noss 1991; Orrock et al. 2000; MacFaden and Capen 2002; Trani 2002; Van Horne 2002).

© The Author(s) 2014
J.K. Keller and C.R. Smith, *Improving GIS-based Wildlife-Habitat Analysis*,
SpringerBriefs in Ecology, DOI 10.1007/978-3-319-09608-7_3

Among many aspects of functional optimization, some such as foraging efficiency, territory size, and territory shape are clearly driven by energetic relationships (Stenger 1958; Schoener 1968; Covich 1976; Blake and Hoppes 1986), whereas others such as predator avoidance and territory placement within the broader landscape may secondarily reflect energetic efficiency (Martin 1992; Suarez et al. 1997; Bakermans and Rodewald 2006). This suggests that the most important explanatory (habitat) variables should be those that most precisely describe the structural and/or energetic aspects of species-specific landscape components and best characterize landscape attributes (i.e., size, configuration, and context) associated with optimal energetics of a species' territory or the biotope in which it occurs (e.g., Wolters et al. 2006). The challenge then is to (1) identify taxon-specific landscape component subsets within biotopes, (2) effectively quantify those components at a resolution and scale appropriate to the species, guild or assemblage of interest, and (3) characterize component spatial distribution and that of the surrounding landscape from the perspective of energetic efficiency and reproductive fitness (see Wiens 1989b for a discussion).

Although the desirability of selecting explanatory variables from the viewpoint of the organism is widely agreed upon, Van Horne (2002) suggested that doing so, in practice, is exceedingly difficult and thus frequently results in selection of variables that "we can measure easily." We agree but submit that if the goal is to characterize observed species–habitat association and/or use and existing variables inadequately do so, it may fall to the researcher to identify more ecologically meaningful variables at a taxon-specific resolution and develop the procedures to quantify them (cf. introduction of Farrell et al. 2013). To us, developing new and biologically intuitive variables that help "connect the dots" between observed species function and habitat characteristics is one of the most exciting and rewarding aspects of field research and is essential for continued advancement in species management and conservation.

3.1 Ground-Based Variables

Our focus is on the informed use of GIS-based metrics derived from remotely sensed imagery in species–habitat modeling and management. However, the relevance of any predictor variable to the function and habitat selection of a species is important for ground-based variables as well. Basal area, bole diameter, and foliage volume, for example, are surrogates for tree size and the presence of high canopy and are frequently included as explanatory variables for forest-associated species. Yet, for many such species, these variables are not as specific as some alternatives. For high-canopy-gleaning birds, leaf area above 7 m is a more function-related and better predictor of occurrence and density than the former variables (Keller 1986; Keller et al. 2003). Similarly, snag density is sometimes a functionally more useful addition to models for predicting woodpecker richness and particularly abundance (Holmes et al. 1979; Keller 1986; Keller et al. 2003) than the previously noted surrogates of tree size. Although basal area, for example, may suggest that tree size is not limiting, snag density, and thus, feeding site and nest site availability, may be limiting. Both examples illustrate the value of quantifying a variable more closely related to the species' function, which in turn dictates habitat selection.

Until recently, higher species–habitat correlations for smaller, less wide-ranging species were more frequently achieved via measurements of habitat structure on the ground, employing measures such as those of James and Shugart (1970) (cf. Krawchuk and Taylor 2003; Hallworth et al. 2008; but see Saab 1999; Bakermans and Rodewald 2006). This is because of the limitations of image resolution and/or variable selection in many studies incorporating remotely sensed imagery. Measurements of landscape components obtained from ground-based samples typically are collected (1) at the finer habitat selection scales (higher resolution) used by smaller species (e.g., passerines, small mammals, insects), (2) in connection with specific habitat subunits such as nest sites of wider ranging species (e.g., Collins et al. 2009), or (3) as the local scale component of studies addressing the influence of multiple scales of landscape structure on species richness (e.g., Ewers et al. 2007) (However, see Chap. 4 regarding potential errors introduced by inappropriately located vegetation samples). The recent adoption of H-resolution light detection and ranging (LiDAR) for quantification of variables associated with solid patches and/or terrestrial species with relatively small home ranges (e.g., less than 3–5 ha) has greatly improved the explanatory power of GIS-based models for such species (e.g., Goetz et al. 2010; Farrell et al. 2013).

Lastly, the ability to separate incremental, rapidly changing (e.g., successional) differences in habitat structure also was, until recently, better accomplished with ground-based metrics such as Aber's (1979) camera technique for foliage profile determination. As noted above, LiDAR (see Zimble et al. 2003; Vierling et al. 2008) now offers a desirable, less labor-intensive alternative to this and other approaches for quantifying vertical subcanopy structure and vegetation density. Variables based on LiDAR such as the Normalized Difference Vegetation Index (NDVI) (Kerr and Ostrovsky 2003) offer potentially effective remotely sensed alternatives to ground-based measures of plant productivity such as leaf area distribution (cf. Keller et al. 2003).

3.2 Landscape Metrics: Size, Configuration, and Context

Just as certain aspects of habitat can be more appropriately quantified on the ground, the planar two-dimensional or spatially more extensive attributes of landscapes such as the size, configuration, insularity, and context of particular landscape elements or plant communities can be more thoroughly quantified using remotely sensed imagery (Keller et al. 1979a, b, 1980; Keller 1986; Forman and Godron 1986; Turner 1989; Turner and Gardner 1991; but cf. Krawchuk and Taylor 2003). However, many landscape metrics either (1) are imprecise measures of actual habitat size and/or spatial arrangement, (2) are applied to habitat or landscape structure unrelated to taxon-specific aspects of habitat selection, (3) fail to consider energetics of habitat use, (4) are measured at image resolutions or geographical scales inappropriate to the taxon of interest and/or (5) do not account for classification errors or other forms of uncertainty introduced during map generation (see Wagner and Fortin 2005 and references therein for additional discussion of problems with landscape metrics). We addressed the critical issues of resolution, analytical scale, and classification errors in Chap. 2 and shall consider the first three issues identified above in this chapter.

3.2.1 Proportions and Indices: Relative Versus Absolute Measures

3.2.1.1 Problems with Proportions

Proportions of various plant communities, land use types, or distributions of a particular habitat component (e.g., canopy cover) are often included as explanatory variables in habitat analyses (cf. Freitas and Rodrigues 2012) but suffer from several limitations (Keller et al. 1979b, 1980). First, they fail to quantify actual area, which is a critical determinant of habitat suitability for threshold or small populations. Second, and more importantly, proportions fail to quantify spatial arrangement. Figure 3.1 illustrates how different spatial arrangements or shapes can yield identical proportions or index values, and thus, how either proportions of cover types or indices of edge density, including fractals, would fail to quantitatively separate different spatial distributions of identical amounts of landscape components.

By inference, use of proportions means that a fixed area is being sampled. Fixed area sampling may be appropriate if the goal is to (1) map individual territories and compare attributes of used versus unused areas within the fixed sample

Fig. 3.1 Hypothetical distributions of identical amounts of two different landscape component types on two plots. Neither proportions, patch density, Plot Edge Density, edge indices, nor fractals would separate these different spatial arrangements of identical amounts of components (from Keller et al. 1979b)

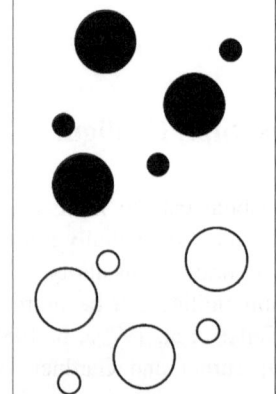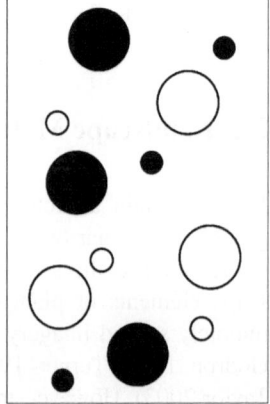

(e.g., Wiens et al. 1987; Aebischer et al. 1993) or (2) examine the regional context in which a particular preidentified habitat or biotope exists (e.g., Pearman 2002; Bakermans and Rodewald 2006). Use of proportions also may not be problematic in riparian or right-of-way (ROW) corridors. Due to their linearity, the variability (range of values) of configurations of a particular proportion is limited, and thus, proportion may provide relatively useful characterization of availability of a measured cover type structured by a hydrologic or anthropogenically influenced gradient (e.g., Saab 1999; Freitas and Rodrigues 2012; but see Miller et al. 2004). Studies of nonlinear landscapes, however, often illustrate the inability of proportions to add much explanatory power to species–habitat models when similar proportions may occur in differing arrangements (e.g., Hiebeler 2000; Donovan and Flather 2002; MacFaden and Capen 2002; Davis et al. 2007; Kays et al. 2008; Betts et al. 2010).

If the focus is on the habitat or biotope itself, we suggest that the landscape of interest be scanned to determine the actual amount of any hypothesized habitat available within it and discuss this technique in Sect. 3.2.3. Despite these problems, proportions are attractive because when spatial arrangement and sample area size are at least partially accounted for in the sampling design (e.g., proportions compared over a range of concentric sample sizes within a landscape mosaic), they can be useful in explaining species composition, richness, differences in territory size, or the influence of landscape context (e.g., Pearman 2002; Leonard et al. 2008; Thompson et al. 2012).

3.2.1.2 Indices

A variety of indices have been developed to describe attributes of complex landscapes. Among these, the most common are (solid) patch or edge *density* indices and indices of *shape*, including fractals (e.g., O'Neill et al. 1988; Riitters et al. 1995; Potvin et al. 2001; McGarigal et al. 2002; MacFaden and Capen 2002; Angel et al. 2010). There are many variations on these basic metrics.

The main problem with index-type variables is that they reduce information to such a degree that many possible configurations of landscape components or cover types can produce the same or very similar values, which greatly reduces their explanatory and, ultimately, predictive utility (Cale and Hobbs 1994; Neel et al. 2004; Vogt et al. 2007). For example, patch density can be defined as the total number of clusters or polygons (i.e., solid patches) of a designated landscape component or cover type (or types) per unit area (McGarigal and Marks 1995). Regardless of the definition of a cluster or the GIS type, raster or vector, the minimum patch density for a single component or cover type in the landscape is one divided by the sampled area. However, if that single cluster is extensive in size (e.g., a forest in an agricultural setting) in an N-celled or N-unit sampling area, there may well be more than N ways, even without regard to orientation or cluster location, for that *single* cluster to be configured. Yet, all of these configurations would produce the *same* patch density index $= 1/N$. This is also true for any particular number of multi-minimum mapping unit clusters, regardless of size distribution or configuration, and is also true when considering the patch density of multiple component/cover types. Clearly, such an index has limited discriminatory power.

Additionally, more information is lost if all component/cover types are considered equal (i.e., to have the same habitat utility to the focal species or taxon) (e.g., Haslem and Bennett 2008). In this case, little management insight is gained even if the patch density of components, undifferentiated by type, is correlated with the richness or density of a focal taxon. As a manager, one would still be left with the question, "Which of the component types (and in what proportions) should be manipulated to produce the desired effect on the resource?" Similarly, conservation planners gain little guidance from such metrics in their attempts to prioritize areas for protection.

Edge Indices

A number of studies have compared edge indices to species abundances or species richness with various degrees of success (e.g., McGarigal and McComb 1995; Saab 1999; Drapeau et al. 2000; Cushman and McGarigal 2002; Thompson and McGarigal 2002; MacFaden and Capen 2002; Betts et al. 2006; Davis et al. 2007; Brennan and Schnell 2007; Schlossberg and King 2008). Edge indices are typically calculated as the length or number of edges observed within a defined area divided by the maximum length or number that could occur there. This may include all edges within the landscape, all edges associated with a particular plant community (solid patch) type, or other combinations. There are many variations on the approach (cf. MaGarigal and McComb 1995).

As noted earlier, edge is frequently used in the literature to refer either to (1) ecotones only (i.e., the edge between two adjacent plant communities) or (2) to **all** edge **types** collectively, thus implicitly equating the value of all edge types to all species (i.e., edge for rabbits = edge for flycatchers). As with patch density, this latter practice of implicitly equating the habitat value of all edge types represents an a priori data reduction that results in an unnecessary loss of information. Recognizing this, Keller and Anderson (1992) noted that the nature of the adjacent plant community is more important than the existence of an edge per se.

We submit that equating all edges is an often unappreciated and sometimes unacknowledged underlying assumption that simultaneously (1) dilutes the power of correlations of species occurrence or abundance with specific edge types more directly associated with the focal taxon and (2) obscures these relationships. This is one of the reasons that even when edge metrics are significant in species–habitat models, they rarely explain much of the variation (i.e., >10 %) in species occurrence or abundance, even for species generally conceded to be associated with edges [cf. proportion of variance explained by edge-related PCA variables in McGarigal and McComb (1995, Table 6), Saab (1999, Table 4) and individual edge-related metrics in Betts et al. (2006, Table A.1)].

One alternative approach that reduces the loss of information associated with edge type aggregation is to weight edges by degree of contrast in height (e.g., seral age) or other major difference between adjacent land use types such as coniferous adjacent to deciduous vegetation (FRAGSTATS, McGarigal and Marks 1995). However, this classification/quantification approach still has at least three limitations. First, the specific components of the edge are lost. Second, weighting the degree of edge contrast is subjective and analyses may be sensitive to the introduction of such arbitrary ratings (e.g., how does one distinguish a 0.5 from a 0.6?). Third, the contrasts, which may vary widely, even along the perimeter of a single cluster or polygon due to the juxtaposition of different patch (landscape component or plant community) types, are averaged for the entire patch type, again obscuring the potential significance of certain component pairings along the cluster perimeter. This occurs because the metric is compiled from the perspective of the particular patch type (i.e., the individual *solid* clusters), rather than the *edge type(s)* in question. Even if the contrasts only were maintained categorically as percentages of total edge length, this would maintain more information about the edge types than the edge index, which in essence calculates a landscape-wide metric that sums the average contrast for each patch across the entire area of interest.

We readily acknowledge that some forest edge species like the red-tailed hawk are relative edge-type generalists and thus potentially amenable to data reduction techniques such as edge contrasts (see McGarigal and McComb 1995, Table 8). However, describing edges simply as having a 0–100 contrast (actually 0.0–1.0 multiplied by 100) based on the structural or vegetation disparity between the adjacent plant communities can result in a loss of much useful information (see examples in Chap. 5) by masking both the occurrence and spatial distribution of potentially critical edge types.

The use of edge types more specific to particular species or guilds has generally yielded better results (e.g., Rehm and Baldassarre 2007; Wilson and Watts 2008). For example, using a more specific, function-related edge type as a potential correlate of species occurrence, in this case deciduous forest-stream edge for Acadian Flycatcher *Empidonax virescens* (ACFL) habitat, Chapa-Vargas and Robinson (2007) found a disproportionately high density of ACFL nesting within 5–50 m of this edge type as opposed to other identified edge types in the study. Edge indices that aggregate types or even those based on contrasts would not have adequately quantified the flycatcher's association with this specific (i.e., forest-stream) edge type (Chapa-Vargas, personal communication; Carpenter et al. 2011). We provide a detailed analysis of edge-specific associations of several early successional bird species in Chap. 6.

Edge Density

Some of the same problems noted above apply to measures of edge density, which is defined as length of edge per unit area, typically km/km^2 or m/ha. As with indices of edge, all types of edge are usually combined or at best viewed as contrasts (but see Rehm and Baldasarre 2007). Again, if included in species–habitat models as edge density of all types combined, or even as contrast-weighted edge, this variable rarely explains much of the variation in species occurrence or abundance (cf. Penhollow and Stauffer 2000; Johnson et al. 2002; Lichstein et al. 2002a; Betts et al. 2006; Whitaker et al. 2007; Haslem and Bennett 2008; Lapin et al. 2013). Brennan and Schnell (2007) found bird density correlations (r) with edge density (combined with fractal dimension) in the $r = 0.25$–0.4 range for the suite of flycatchers they studied. The senior author found similarly modest correlations of bird species' presence and density with generalized combinations of edge (Keller, unpublished) in analyses of species-edge associations (Keller 1986, 1990). He then found, however, significantly higher correlations and produced highly predictive stepwise discriminant function analysis (SDFA) models (op. cit.) using more specific edge types (cf. Fig. 1.2) such as described in the preceding section on edge indices and in Chap. 6.

Fractals/Shape Complexity

Similarly, fractals, another form of edge index, are frequently used to measure shape complexity but again, typically, for all edge types combined (e.g., McGarigal and McComb 1995; Saab 1999; Johnson et al. 2002; Betts et al. 2006; Brennan and Schnell 2007; Haslem and Bennett 2008). Correlations of species occurrence or richness with these generalized edge measures are similarly often low or nonsignificant (e.g., Lawler et al. 2004; Lapin et al. 2013). In addition to the loss of information due to edge-type aggregation, the scale of analysis again is likely problematic.

Shape

Like the edge measures above, shape is most often quantified as an index, in this case with a value relative to that of a circle (=1), which has the minimum perimeter to area ratio of any shape. Applied to describe to solid patches, shape is frequently defined as the sum (either for a single patch, or at the landscape level for all patches, of type T) of the length of the patch perimeter divided by the square root of patch area, corrected by a constant including pi so that a circular patch has a value of 1. However, because patches of very different sizes can produce similar shape indices, calculating shape without considering its interaction with size can obscure any effect of shape even when shape is measured for individual patches.

In addition, as with the measures discussed above, averaging shapes for all patches of a given type has the effect of masking any trend in shape change across that patch type. Thus, by itself, shape is only occasionally useful as an explanatory variable, as evidenced by its infrequent inclusion in habitat models either for individual species or higher organizational levels (e.g., McGarigal and McComb 1995; Cushman and McGarigal 2003; Lawler et al. 2004; Lapin et al. 2013). Even when included, results are frequently equivocal (e.g., Betts et al. 2007).

More recently, morphological spatial pattern analysis, now included in the free online software toolbox GUIDOS (http://forest.jrc.ec.europa.eu/download/software/guidos), provides user-specified options to classify different spatial arrangements of identified cover types into "shape" categories generally recognized to influence species distributions and movement (Vogt et al. 2007; Ostapowicz et al. 2008). It is important to recognize, however, that terms such as "core," "island," "isthmus," and "peninsula," are only relative and their applicability is entirely taxon specific. For example, what constitutes a "corridor" to a deer may be "core habitat" to a shrew, but would be misclassified as the former shape for the shrew at Landsat resolution. Second, although cell size (P) in this application is user defined, both its lower size limit (i.e., minimum size delineation) and resolution are fixed a priori by the original image source, points not discussed in either of the preceding articles.

Additionally, the choice of two different ways to calculate edges offered by GUIDOS illustrates one of the problems of square pixels in attempting to quantify edge. One option considers edges as the traditional four linear interfaces shared by a given cell with adjacent cells. The second option additionally considers edges to include the point interfaces shared by diagonally adjacent cells, rendering them equivalent to cells sharing a linear interface (see Keller et al. 1980, Figs. 3 and 4). As a result, this choice of what constitutes an edge can produce different estimates of both the absolute and relative amounts of individual edge types for the same image. Keller et al. (1979b, 1980), Burt (1980), and Star and Estes (1990) previously discussed the advantages of hexagonal versus square pixels for quantification of edge and spatial arrangement.

Conclusions for Sect. 3.1–3.2.1:

- Variables based on *proportions* do not consider actual spatial distribution, habitat size, or shape, but can be useful in assessing (1) species–habitat relationships within corridors or fixed size comparative samples and (2) effects of spatial context.
- Proportions and many indices fail to address energetics, which is important in determining both threshold habitat size and the size of biotopes necessary to support metapopulations.
- Variables, including indices, that combine all landscape component, cover or edge types, or consider merely contrasts between adjacent component types (1) result in an a priori loss of information, (2) are less explicit and informative than species- or taxon-specific types, and (3) can introduce unnecessary "noise" into the analysis. This can result in lower correlations with the focal taxon's distribution and abundance and ultimately lead to false interpretations of species–habitat relationships.
- Identifying specific landscape component types or edge types that are meaningful to particular species or guilds appears critical to improving the predictive power of species–habitat models (see Chap. 6).
- Lastly, when generalized indices, especially less intuitive ones such as fractal dimensions, are included in models that employ canonical correlation (CCA, PCA) and similar combinatoric techniques, the difficulty of interpreting the results and developing management recommendations based on such

variables is increased (see Appendix B). Although seen infrequently, some authors (e.g., James 1971; Murkin et al. 1997) even have included pictograms in an attempt to illustrate what these multivariate axes actually mean.

3.2.2 Size

MacArthur and Wilson's (1967) seminal reduction of observed species richness patterns on oceanic islands into the now fundamental island biogeography theory has spawned countless studies of the influence of biotope size on species occurrence. Many papers have followed arguing for and against effects of island size on various taxa in diverse settings. In many cases where biotope size was not well correlated with species richness or composition, alternative explanations for richness were posited or tested. Among alternative explanations, three appear frequently.

- Internal heterogeneity—often a function of size but confounds effect of size alone [see Gilbert 1980 for a review, Litwin and Smith (1992), and Matias et al. 2010 for a more current discussion]
- Variability in the shape of the "islands," which implies that not all shapes are equally colonized (Osman 1977)
- Composition of the surrounding landscape (e.g., Litwin and Smith 1992; Saab 1999; Rodewald and Yahner 2001; Pearman 2002; Kennedy et al. 2011)

Despite the sometimes conflicting evidence for size effects, several underlying principals related to the importance of size alone are consistent on both habitat islands and oceanic islands:

- First, larger species occur only in larger biotopes (habitat islands). This is a direct result of the relationship between increasing body size and increasing territory size (Schoener 1968).
- Second, larger patches, solid or edge, of appropriate habitat should contain more guild members because a patch that is suitable for a large member of a guild should more likely be suitable for a smaller member of that guild also (Keller 1986).

Therefore, the size of patches is still worth considering, and we suggest that one of the reasons that size is not consistently correlated with species richness is the way it is measured. Specifically, *measures of size do not include shape*. Throughout the literature, when either variable is considered, it generally is measured and considered independently of the other (but see "landscape composition" variable in Collier et al. 2012).

3.2.3 Considering Energetics: The Basis of Optimal Patch Shape

The importance of habitat shape (more typically plant community shape) to a species occurrence on the landscape has been studied extensively with conflicting results (for a discussion, see Flather and Bevers 2002). Various measures of habitat arrangement or shape have frequently been found to have little to no effect on species occurrence, abundance, or richness (e.g., McGarigal and McComb 1995; Meyer et al. 1998; Drapeau et al. 2000; Cushman and McGarigal 2003; Haslem and Bennett 2008). Yet, arrangement also has been found to be important, both theoretically and empirically, at patch sizes that include small populations (i.e., on islands, whether "habitat" or oceanic) (Simberloff 1976; Keller 1986, 1990; Andren 1994, 1996; Schumaker 1996; Fahrig 1997, 1998; Hiebeler 2000; Flather and Bevers 2002; Cushman and McGarigal 2004; Davis 2004). Hiebeler (2000), for example, found populations on simulated landscapes were largely determined by the degree of habitat clustering, while habitat amount had little effect. Flather and Bevers (2002) expressed the confounding nature of the evidence on this issue. While suggesting there was no point in searching for an optimal arrangement, they simultaneously warned that shape should not be ignored because it is likely important at or near threshold population levels.

Among the studies cited above that found a relationship of population size with patch shape, **clustered** arrangements of habitat were consistently cited as being important. If so, what attribute of clustering produces this consistent observation? We suggest energetic efficiency of clustered arrangements versus random, linear, regular, or other types of spatial distributions is the basis for their criticality to threshold or small population sizes. Specifically, the fact that for small population sizes, clustered arrangements reflect a relationship between optimal habitat or patch shape and optimal territory shape.

Covich (1976) argued that, in an energetics context, a circle is the optimal shape in horizontal space for an all-purpose territory. This is the result primarily of two properties of circles: (1) A circle has the minimum perimeter to area ratio; therefore, it has the minimum amount of border per unit area to defend against conspecifics, and (2) it has the shortest average distance between points within it, which optimizes foraging.

3.2.3.1 Quantifying Size, Shape, and Composition Simultaneously

Few, if any, landscape metrics directly consider energetically optimal habitat shape from this species-centric perspective, and no variables that we reviewed consider size, optimal shape, and habitat composition simultaneously (see Flather and Bevers 2002 for a theoretical discussion). To address this issue, Keller (1986) developed a two-dimensional descriptor of patch clumping and potential territory size based on the arguments of Covich (1976) that a circle is the optimal shape in horizontal space for an all-purpose territory. Applied in GIS to solid patch types such as forests and open grasslands, this variable is the maximum diameter circle (MDC) that fits within any solid species- or taxon-specific structural type identified in the imagery (Fig. 3.2). MDC thus represents the *functional size* of the taxon-specific solid patch regardless of its actual size and shape.

In a discussion of moving window analysis in GIS, Wagner and Fortin (2005: 1984; see also Bradshaw and Fortin 2000) noted that such analyses use windows (i.e., scanning fixed area samples) *"of arbitrary size that do not reflect the spatial structure of the species or the environment"* and that *"research in geographical information sciences should address this issue in order to provide tools for detecting the patchiness or zone of influence of the data ...and implementing flexible geographical (e.g., watershed) or behavioral (e.g., home range) windows that can be adapted to a specific situation."* As depicted in Fig. 3.2, we submit that MDC, by virtue of its representation of the functional size of a patch, or biotope (i.e., it reflects the structure of the environment and is organism based) provides such a flexible, behaviorally based "window" for detecting the *"patchiness or zone of influence of the data"* sought by Wagner and Fortin (2005) (see also Sect. 4.2).

Among examples of unrecognized but similar applications of this approach, Keller et al. (1993), Darveau et al. (1995), Hodges and Krementz (1996), Burhans and Thompson (1999), Confer and Pascoe (2003), King et al. 2009b (all bird species), Stoddard and Hayes (2005; salamanders), and Kubel and Yahner (2008; Golden-winged Warbler) all found species richness correlations or threshold occurrence relationships with the width (=MDC) of various corridor types (e.g., powerline ROW, fencerow, riparian corridor). Davis (2004) found that among similarly sized native grassland patches, those with a lower perimeter (edge) to patch area ratio (i.e., a larger MDC) supported greater richness and abundance of area sensitive grassland birds.

Figure 3.3 illustrates the application of MDC's to various structural types identifiable in an aerial photograph. However, note how heterogeneity of the oldfield example at the upper right does not fit the definition of a "solid" patch type. The question immediately arises as to how to apply the MDC approach to quantifying patches for edge species? To measure the size of edge patches (e.g., shrub-open grass edge and shrub-water edge) in a manner comparable to MDC, Keller (1986) also developed an edge-scanning algorithm (ESCAN). ESCAN locates areas with the highest density of edges (m/m^2) for a given structural type T within a series of progressively larger circular samples on the GIS map (Fig. 3.4). For Keller's (1986) study, these samples were actually hexagonally shaped, as the GIS he employed (Keller et al. 1979a) was based on hexagonal-celled pixels.

Fig. 3.2 The maximum diameter circle (MDC) representing the largest optimally shaped territory that fits within a particular "solid" patch type located within multiple biotopes identifiable on remotely sensed imagery. MDC represents the functional size of the patch

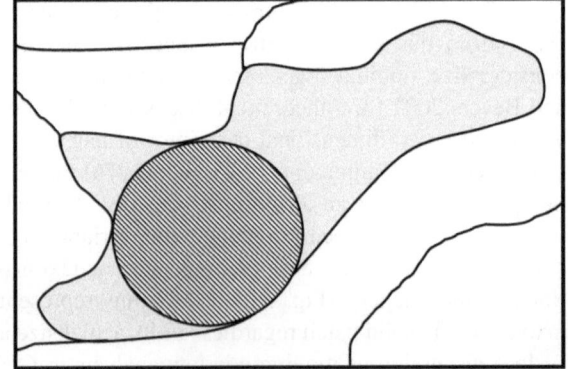

Fig. 3.3 MDC's for different patch (in this case, plant community) types. Note the heterogeneity in the oldfield at the *upper right*, which is inconsistent with the concept of MDC as pertaining to "solid" patch types (see text and Working Definitions)

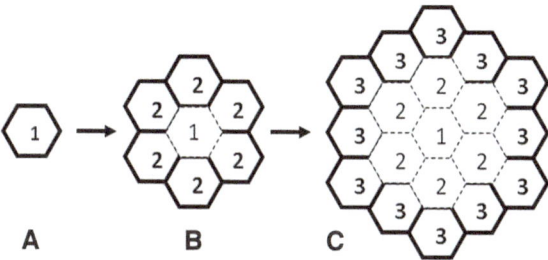

Fig. 3.4 The pattern of edge addition (*solid lines*) to the progressively larger samples examined by the ESCAN algorithm. A prespecified number of annuli are examined starting within each cell on the GIS map. The program locates the area on the map with the highest density of edges of any specified patch type (T) for each sample size (i.e., number of annuli). *A* 1 annulus sample = 6 edges, *B* 2 annuli sample = 30 edges, *C* 3 annuli sample = 72 edges total. See text and Table 3.1

Using ESCAN, Keller (1986) was able to calculate a variable equivalent to MDC called the diameter of the equivalent area circle (DEAC) for all structural types composed of edges (Fig. 3.5 and Table 3.1; see Keller 1986, 1990 for discussion). In addition to the ecologically intuitive properties of MDC described above, we suggest the DEAC resulting from the systematic, exploding scan sampling

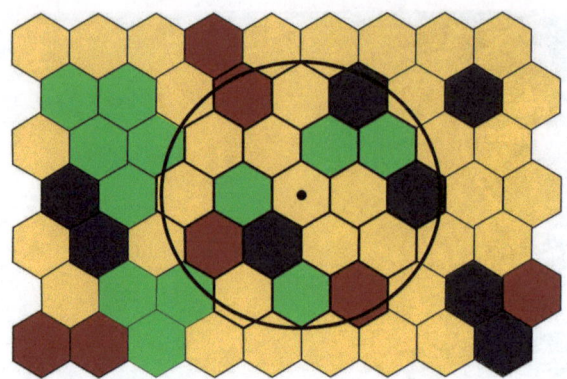

Fig. 3.5 An example of the DEAC resulting from an ESCAN analysis of a hypothetical old-field. In this example, the program locates the highest density (m/m^2) of shrub–sapling/opening (=Type T) edges on the plot and translates the area sampled into an equivalently sized *circle* (DEAC) comparable (for edges) to a MDC for "solid" patch types (see ESCAN program output in Table 3.1). Note, for illustrative purposes, all plot perimeter edge is non-Type T (i.e., all Type T is internal). In practice, edge between the plot and adjacent cover types also could contain Type T edges. Black cells = coniferous trees, sienna cells = deciduous trees, green cells = deciduous shrubs, tan cells = grass

Table 3.1 An example of the output from the edge density scanning algorithm ESCAN as applied to shrub–sapling/opening edge within the hypothetical oldfield in Fig. 3.5

Annulus[a] Size	Starting cell[b]		# Type T	Total edges	% T	AAS[c]	EI[d]	DEAC[e]
	I	J						
1	2	8	6	6	1.000	1	6.00	
2	3	7	18	30	0.600	7	6.80	
3	4	5	43	72	0.597	19	9.87	49.19
4	4	5	58	132	0.439	37	9.54	

See text and Figs. 3.4 and 3.5 for further description of the technique
[a] Number of annuli (i.e., rings of cells) in the sample
[b] Location of the row (I) and column (J) of the sample center on the hexagonal-celled map depicted in Fig. 3.5
[c] The actual area (# of cells) sampled by the procedure
[d] Modification of Patton's edge index: EI = No. of Type T edges in the sample/AAS$^{.5}$
[e] Diameter of the equivalent area circle (m) calculated only for the annulus size >1 with the highest EI

approach of ESCAN has interpretational simplicity when compared with edge metrics either (1) based on fixed-point samples (e.g., Sisk et al. 1997), (2) fixed area samples (i.e., "windows"; cf. Riitters et al. 2000; Potvin et al. 2001), or (3) compiled for entire (fixed area) landscapes (e.g., McGarigal et al. 2002).

As an example of the discriminatory power of MDC and DEAC, Donovan and Flather (2002) attempted to classify large-scale landscapes as fragmented or unfragmented using several GIS metrics (e.g., shape indices and forest area) and offered two examples of landscapes that were misclassified using these variables (see Fig. 2c and d therein). Selection of the larger of the values of MDC or DEAC calculated for solid forests (unfragmented) and forest edges (fragmented), respectively, on these figures correctly classifies both misclassified images as "fragmented" or "unfragmented" (Keller, unpublished). More recently, Vogt (2013) stated that "...*fragmentation is usually defined from a species point of view but a generic and quantifiable indicator is needed to measure fragmentation and its changes.*" We submit that MDC represents such an indicator. In other examples, use of DEAC allows identification of both the functional size and specific location of "high-quality habitat" within the mosaic of habitat quality depicted in Fig. 1(c) of Franklin et al. (2002) and identifies the locations of high-density populations associated with juxtaposed patch types in Figs. 1 and 2 of Dunning et al.'s (1992) discussion of landscape complementation and supplementation (Keller, unpublished). One also can easily observe that MDC correctly classifies the high, medium, and low probability habitats for Connecticut Warbler (*Oporornis agilis*) shown in Fig. 4 of Lapin et al. (2013).

Lastly, Kotliar and Wiens (1990) identified three criteria, (1) scale, (2) contrast [i.e., component types], and (3) aggregation [i.e., spatial arrangement], as necessary for implementation of a hierarchical approach to the study of patch heterogeneity. MDC and DEAC address all three criteria and, unlike many metrics, which are unstable and difficult to interpret at extreme values of aggregation or the proportion of a given patch type (Neel et al. 2004), are consistent because they are based on two underlying biological relationships—(1) the circle as energetically optimal territory shape and (2) the correlation of body size with territory size. These latter points also address Van Horne's (2002: 71) criticism that "*spatially explicit analyses tend to be highly empirical and have relatively little theoretical foundation.*"

What Do the Circles Mean?

- If two patches of the same type are compared, the one with the larger MDC or DEAC should typically accommodate both larger species (i.e., with larger territories) and more species (i.e., increased species richness) within a guild associated with that patch type.
- This is true until the area containing the patch type is multiple times the diameter of the largest territory of interest (i.e., until metapopulations occur at the landscape scale).
- At that point, territory shape and locations relative to one another are unconstrained by the shape of the available patch type, and patch shape becomes unimportant (Simberloff 1986; Andren 1994, 1996; Fahrig 1997; Flather and Bevers 2002). This explains the lack of correlation of population size with patch *shape* at large population sizes, where patch *size* alone may be important.

Conclusions for Sect. 3.2.2–3.2.3:

- The energetic efficiency of *clustered* arrangements versus random, linear, regular, or other types of spatial distributions is the basis for their criticality to threshold or small population sizes. Specifically, for small population sizes, clustered arrangements reflect a relationship between optimal habitat or patch shape and optimal territory shape.

- *Shape indices* by themselves are rarely meaningful, but can be more biologically informative and interpretable when (1) considered from an energetics perspective and (2) combined with metrics that quantify the size of species- or taxon-specific patch types, whether solid or edge.
- Similarly, measures of habitat or patch *size* that do not consider energetically optimal arrangements of species- or taxon-specific landscape component types may result in misinterpretations of the influence of patch size on species occurrence or richness.

3.2.4 Configuration

3.2.4.1 Measures of Spatial Arrangement

Many studies purport to measure spatial arrangement but may capture only limited aspects of landscape component configuration because they predominantly employ variables (e.g., proportions and indices) for which identical values can be obtained for very different configurations (e.g., Table 2 in Lichstein et al. 2002a, Table 2 in Drapeau et al. 2002; and see Sect. 3.2.1). Researchers generally agree that combinations of variables that quantify different aspects of spatial arrangement are required to describe more fully the configuration of landscape components in a GIS scene (e.g., Wagner and Fortin 2005). Techniques such as gradient analysis (Keller et al. 1979b, 1980, Fig. 3.6), nearest neighbor analysis, the runs test (see Chap. 6),

Y axis \ X axis	1	2	3	4	5	6
1	A	A	A	B	B	A
2	A	A	B	B	A	B
3	A	B	B	A	B	B
4	B	A	B	A	C	C
5	A	A	A	A	A	C
6	A	C	A	C	C	A
A	5	4	3	3	2	2

Fig. 3.6 An example of application of spatial analytical techniques such as the runs test to a hypothetical GIS map with three landscape component types, *A*, *B*, and *C*. Analysis for a trend in the proportion of Type *A* across the map from left to right can be accomplished by summing the values of *A* in each column (i.e., along the *Y* axis) and then statistically testing the resulting univariate data for a trend along the *X* axis, as indicated by the *arrow* (from Keller et al. 1979b)

circumcircle methods, the Local Index of Spatial Association (LISA), and spatial autocorrelation coefficients such as Moran's I can be used to separate random, clustered, or regular distributions mathematically (see Dale et al. 2002; Fortin et al. 2002 for reviews). Additional methods such as high-resolution image texture analysis (St. Louis, et al. 2006) can quantify other aspects of spatial heterogeneity or pattern.

As discussed in Sect. 3.2.3, we suggest that MDC and DEAC represent more organism-based measures of configuration, in this case the degree of clustering of specified landscape components (solid or edge) that can be tested against the distribution and abundance of a particular species, guild, or assemblage. Although a discussion of spatial analytical techniques is not the focus of this paper, the use of these techniques in conjunction with other GIS metrics is critical to properly assess the influence of landscape component spatial arrangement on species' distributions and abundance.

Based on the foregoing discussion, we recommend that researchers:

• Identify landscape component subsets (solid or edge types) specific to the focal taxon [see recommendations (p. 464) regarding variable selection by James and McCulloch 2002],
• Use multiple variables that quantify different aspects of spatial arrangement to separate landscapes that may be perceived as different by wildlife but that might statistically appear the same if measured with one or a few variables that can produce identical values for different configurations, and
• Include variables such as MDC and DEAC (see Sect. 3.2.3) that quantify the degree of aggregation of taxon-specific landscape components while simultaneously considering size and optimal shape.

3.2.5 Context

With the incorporation of remotely sensed imagery and GIS into community ecology, researchers began to recognize the potential importance of surrounding landscapes (i.e., the "matrix") on population dynamics (Wiens 1989b; Forman 1995; Freemark et al. 1995). Although in reviewing research prior to 1999, Mazarolle and Villard (1999) found local variables to be more important than landscape variables, dozens of studies since have found significant correlations between the composition of a site's surroundings and its species richness or use by particular species (e.g., Saab 1999; Drapeau et al. 2000; Johnson et al. 2002; Lee et al. 2002; Pearman 2002; Bakermans and Rodewald 2006; Pillsbury and Miller 2008; Kennedy et al. 2011; Thompson et al. 2012).

Similarly, many studies have examined the influence of landscape composition on brood parasitism, nest depredation, and various measures of nest success (e.g., Donovan et al. 1997; Rodewald and Yahner 2001a; Thompson et al. 2002; Driscoll and Donovan 2004; Winter et al. 2006). Among these studies, the size of the matrix analyzed varied widely from circles of 10s of meters in diameter to 5 km or more. The level at which context was found to influence local species occurrence or reproductive success, if at all, also varied widely depending on the focal taxon, the biotope examined, and the structure/composition of the surrounding land use (see discussion by Kennedy et al. 2011). In general, the most structurally different contexts, particularly anthropogenically influenced settings (e.g., forests in suburbia), had the greatest effect on the focal taxa (e.g., Bakermans and Rodewald 2006; Kennedy et al. 2011), particularly less mobile ones, such as amphibians (Pillsbury and Miller 2008; reviewed by Mazerolle et al. 2005).

3.2.5.1 Other Taxa

Lastly, while our discussion has focused on birds, the principles under discussion apply to many other taxa such as amphibians referenced above (cf. Stoddard and Hayes 2005; McKenny et al. 2006) and small mammals. The red-backed vole (*Clethrionomys gapperi*), for example, is often referred to as a mature northern hardwoods–hemlock (NHH) species, but its densities have been found to be higher on young clear-cuts than in adjacent mature forests (Kirkland 1977; Keller unpublished data). As with our common yellowthroat example earlier, (1) the presence of the correct landscape components [e.g., mesic, closed canopy NHH forests with abundant forage and coarse woody debris on the ground, regardless of forest age], (2) at or above a threshold size [partially dictated by the composition of the surrounding landscape], with (3) optimal shape, yields appropriate habitat. Denser, more productive contexts, such as young clear-cuts, yield the highest densities (Kirkland 1977; Keller unpublished data).

Similarly, remaining colonies of the Allegheny Woodrat, whose isolated populations are disappearing throughout its range, have been noted to be associated with the largest remaining areas of highest density talus-forest edge (i.e., the variable DEAC discussed in Sect. 3.2.3) in the Ridge and Valley Physiographic Province and along the Alleghany Front in the eastern U.S. (Cal Butchkowsky, PA Game Commission, personal communication).

Chapter 4
Landscape Sampling Areas Versus Actual Location of Taxonomic Survey

4.1 Point-Count Assumptions and Inferences

Many bird studies use point-count surveys to develop species distribution and abundance data, which are then correlated with ground-measured habitat data, GIS data, or combinations of the two to produce species–habitat models (e.g., McGarigal and McComb 1995; Saab 1999; Lichstein et al. 2002; McFaden and Capen 2002; Betts et al. 2003, 2006, 2007; Miller et al. 2004; Brennan and Schnell 2007; Smith et al. 2008; Howell et al. 2008; Dickson et al. 2009; Hartman et al. 2009; Graves et al. 2010; LeBrun et al. 2012). Regardless of whether habitat variables are measured on the ground in association with the sample point or whether GIS metrics are quantified for the biotope(s) surrounding the sample point, explanatory variables developed in association with point-count data have one or more underlying assumptions that are rarely, if ever, met or acknowledged (but see Young and Hutto 2002: 109). We shall examine two typical types of analyses based on point-count data to explore the influences of these underlying assumptions.

4.1.1 Local Scale

A major assumption of point-count samples is that for species *recorded* at that sampling point, the *entire area* within the listening/sampling radius surrounding the sampling point is either currently used by (i.e., within the territory of) or represents habitat for the species. Figure 4.1a, b (after Keller et al. 1979b) illustrates how fixed-location or fixed-area habitat samples from within this listening/sampling radius may either characterize species habitat with data from outside the surveyed species' territory or fail to sample fully habitat within the species' territory.

© The Author(s) 2014

J.K. Keller and C.R. Smith, *Improving GIS-based Wildlife-Habitat Analysis*,
SpringerBriefs in Ecology, DOI 10.1007/978-3-319-09608-7_4

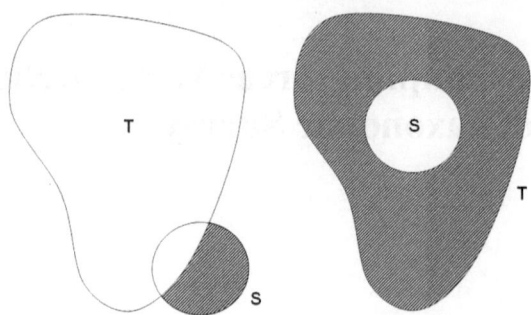

Fig. 4.1 a An often used sampling technique for determining avian–habitat relationships is to randomly select *n* points within the study area; each point represents the center of a sampling unit of predetermined size and shape, in this case, a *circle*. Each unit is sampled for habitat variables and detection of a particular species within the unit is recorded (i.e., a point-count survey). *Shading* depicts the area that is not part of the territory (*T*) of an individual of the species in question but that could be included in a sample plot (*S*) in which the species was observed. This would result in characterization of the species territory by data values that actually were derived from an area **outside** the territory (from Keller et al. 1979b). **b** A second commonly used technique for determining habitat association is to locate, as nearly as possible, the center of an individual's territory or, alternatively, a nest or den site. This point is then the center of a sampling unit of predetermined size and shape. *Shading* depicts the portion of the actual territory (*T*) that remains unsampled by this method (from Keller et al. 1979b)

Samples based on telemetry-estimated or spot-mapped territories (in the case of passerines) largely overcome this objection (e.g., Barg et al. 2005; Wiens et al. 1987, respectively) and Smith (1990) has discussed the various criteria necessary for confirmation of breeding and establishment of a durable habitat association for a species.

The consequences of this point-count assumption of uniform habitat use or suitability may be illustrated best by ground-based samples when deliberately attempting to compare the habitat features of areas where the species is detected with areas where it is not detected (i.e., determining habitat preference). Hartman et al. (2009) studied Cerulean Warblers (CERW) and attempted to differentiate "occupied" from "unoccupied" point-count forest samples in Kentucky. Each sampling station had an assumed effective listening radius of 50 m, which resulted in a sampled area of almost 8,000 m^2 for birds. Vegetation was then sampled from the entire 0.8-ha area rather than from where the birds actually were detected; thus, implicitly assuming the birds were using the entire listening area.

None of the eight models proposed to differentiate occupied from unoccupied habitat were significant, likely due to the inclusion in the "occupied" habitat dataset of many vegetation samples that actually came from unoccupied areas (i.e., they were misclassified). During the discussion of their results, the authors recognized this sampling problem as the likely cause of the lack of significance of the results. Although beyond the scope of our discussion, there also may be an issue of detectability (MacKenzie et al. 2002) with CERW, as with many small passerines (cf. Collier et al. 2012).

This same issue can arise for GIS-based analyses as illustrated in Fig. 3 from Howell et al. (2008). Here, the minimum sampling "home-range-level" units used to quantify habitat (1) may include occurrence data for birds located largely *outside of the unit* sampled for habitat and/or (2) almost invariably will include GIS-derived habitat data from *outside the territory* of a surveyed species due to the size (=5.76 ha) of the minimum habitat sampling unit, which is several times larger than typical territories of the passerines surveyed in the study. This multi-scale study employing Landsat data at a 1:24000 mapping scale concluded that habitat selection was stronger at the "community" (=144 ha area) and higher (3,600–90,000 ha) landscape levels than at the local (i.e., 5.76 ha) level. We ask, would higher resolution imagery with minimum mapping units (MMU) scaled to take advantage of such imagery, samples more closely tied to actual territory locations at the finest scale of analysis, and a minimum habitat sampling unit more closely approximating actual territory sizes, or smaller, produce different results (e.g., better than the 60.5 % correct presence/absence classification rate obtained) and conclusions? See Wiens (2002: 747) for a tangential discussion of these issues.

Similar questions can be asked, for example, of the smallest area sampled in Betts et al. (2010) where the minimum GIS sampling area (150 m radius = 7.1 ha) was nine times as large as the area of the point-count sample for birds (50 m radius = 0.785 ha). In describing their choice of study scale, the authors suggested that the 7.1-ha GIS sampling area with 30-m pixel resolution "captured variation in broadleaf cover at the scale of individual songbird territories" for the group of 12 Neotropical migrant passerines studied. Yet, only 4 of the 12 species had discrimination models for predicting threshold occurrence that could be considered useful in management using the selected criterion for evaluating model success [i.e., area under the receiver-operating characteristic curve (AUC) >0.7].

Inclusion of a spatial autocovariate, in this case, the probability of observing a species at one point conditional on the presence of the same species at neighboring sample points, to account for spatial dependency substantially improved the discriminability of all models. However, as expected, in most cases, the apparent effects of broadleaf cover (i.e., the predictor variable) on bird occurrence were reduced as a result (see our discussion of Spatial Autocorrelation in Sect. 5.4). Although in discussing their results, the authors recognized the coarse resolution of both the vegetation variables (e.g., early broadleaf cover) and mapping scale (there was no mention of image resolution) as likely causes of the reduced explanatory power of the models; they made no mention of the mismatch between vegetation samples and either bird sampling areas or actual territory locations.

As another example of the first type of sampling problem noted above (Fig. 4.1a), Lichstein et al. (2002a) assumed an effective listening radius of 75 m (i.e., nearly 1.8 ha), while sampling local vegetation within only 10 m of the listening post (i.e., 0.03 ha). No overt assumption was made about species use of the vegetation sample; only that this sample, which represented only 2 % of the bird sampling area, was at least characteristic of that area. However, since the regression models developed from the dataset attempted to characterize habitat for the abundance of the 21 bird species studied (essentially comparing characteristics of used vs. unused areas), there is still the inference that if a species was present within the listening radius, the associated vegetation sample was within its territory. Given that most of the species in this study have territories smaller than 1–2 ha, there is no assurance this is the case. This again suggests a possible error of the type illustrated in Fig. 4.1a; see also Smith et al. (2008). The authors recognized this potential problem in discussing the uniformly low regression coefficients obtained for the 21 species studied (i.e., only one species $R^2 > 0.5$; 16 of 21 species $R^2 < 0.25$). They also recognized the potential limitations of the mapping scale (1:24000; no resolution given) and land cover classification employed in discussing the strength of their results. We offer an alternative analysis in Chap. 6 of occurrence and density data for several of the species studied by Lichstein (2002a, b) and Howell et al. (2008).

Determination of habitat use can be equally problematic for wide-ranging species. Unless the territory boundaries are fully known (e.g., via telemetry data), calculation of GIS metrics for areas inferred to be habitat may include areas little used or completely unused by the species of interest. For example, Davis et al. (2007) studying fisher (*Martes pennanti*) habitat in California, compared two extensive areas that included fishers (but without radio telemetry data on home-range usage) with a large area where they were absent. This opens the possibility that although fishers occurred across the occupied region, the GIS metric values associated with their presence actually included samples from unused areas within the region, which would reduce the predictive power, and thus the management utility, of the models. Keller et al. (1979b, Figs. 6–8) suggested a GIS-based method of generating pseudo-territories with random shapes and locations for quantifying habitat use (i.e., where they are vs. where they could be), and many other such procedures now exist in GIS packages.

Lastly, in addition to the assumptions above, both ground- and GIS-based samples may extend assumptions of habitat use or suitability to larger geographic areas by assuming that areas surrounding the point sample are homogeneous (e.g., Smith et al. 2008; see Bart et al. 1995 for a discussion). This is not an issue when the research question directly pertains to site context (e.g., Bakermans and Rodewald 2006). However, if the research question includes the assumption of homogeneity, we demonstrated in the previous chapter that it is frequently untrue, especially when GIS data are based on coarser (i.e., low resolution, large GSD) images.

4.1.2 Biotope and Landscape Scales

At larger geographic scales, areas analyzed via GIS for landscape structure are frequently an order of magnitude (e.g., some species in McGarigal and McComb 1995), or more, larger than either the territories of or the area sampled for bird species of interest (e.g., Betts et al. 2003, 2005, 2006; Miller et al. 2004). This is often true for studies using remotely sensed imagery to quantify landscape characteristics along Breeding Bird Survey (BBS) routes (e.g., Thogmartin et al. 2004b; Brennan and Schnell 2007) and is in addition to the problems noted in Sects. 2.2–2.4 regarding typically coarse (low resolution) imagery, landscape classification systems, and minimum areal units considered for analysis of landscape structure.

For some research questions, such as those investigating the influence of landscape diversity on species richness, workers might intentionally select a range of heterogeneity across landscapes and BBS routes to explore this relationship. However, if the research question pertains to individual species, although the bird and habitat sampling areas may appear homogeneous at the image resolution employed, there can be heterogeneity within both sample types (bird and habitat) and areas analyzed (local and landscape) that is unmeasured, yet critical to species' distributions. For example, Veech et al. (2012: 262) found a *"satisfactorily high-level of similarity in land cover composition between landscapes immediately adjacent to North American BBS routes (buffer distance of 0.4 km on either side of route) and the larger landscapes (buffer distance of 10 km) in which they are embedded."* However, these comparisons were made at a 30-m resolution [grid cells of the National Land Cover Database (NLCD) = 30 m GSD] using only proportions of cover types and three landscape metrics—patch density (see problems associated with this metric in Sect. 3.2.1), largest patch index and aggregation index. As we previously pointed out in Chap. 3 and demonstrate in Chap. 6, this relatively coarse level of resolution coupled with variables that measure only limited aspects of spatial arrangement (i.e., proportions, density, and indices) cannot adequately describe landscape heterogeneity at the finer scales of resolution used by passerines. Additionally, Bart et al. (1995) have discussed the potential errors associated with BBS samples even when the bird and vegetation sampling areas are reasonably matched, and Thogmartin et al. (2004a) have noted the misclassification problems of the NLCD.

In one BBS example, Brennan and Schnell (2007) analyzed habitat associations for eight species of flycatchers along BBS survey routes in the south central USA using 1:40000 aerial photography. They chose a 2.4 km belt width for GIS landscape analysis based on the openness of much of the landscape and on the assumption that the maximum distance at which the widest ranging species, the Scissor-tailed Flycatcher (*Tyrannus forficatus*), was detectable was 1.2 km (Gary Schnell personal communication). This assumed detection radius resulted in a bird sampling area of 452 ha at a single BBS stop. The associated GIS sample at the same stop would be 192 ha (=0.8 km length × 2.4 km belt width). Thus, the potential bird sampling area represents an area more than twice as large as the landscape area analyzed at a given stop and might include birds detected beyond the limits of the landscape sample. Since both the bird sampling and GIS landscape areas far exceed the territory sizes of any of the species in the study, they additionally result in likely mismatches of bird occurrence or density with GIS metrics of the associated landscape sampling areas (Fig. 4.1a).

In a second BBS example, this time using Landsat imagery, Thogmartin et al. (2004b) examined CERW occurrence and habitat associations across a large portion of the species' north-central range. They attempted to construct a geographically extensive spatial model of predicted abundance using several GIS metrics along with latitudinal, terrain, and climatic data at three different scales of analysis (800, 8,000, and 80,000 ha), the smallest of which was the entire BBS transect length (40 km) with a 100-m buffer (i.e., 800 ha = 40 km × 200 m). Unlike the previous flycatcher example (Brennan and Schnell 2007), both GIS data (e.g., proportion of the area in deciduous forest) and bird data were combined for the entire transect length and belt width (i.e., 800 ha). This means that the smallest scale of analysis is almost three orders of magnitude larger than the territory size of the warbler, even without considering the likely smaller (i.e., subterritory) grain size at which it selects habitat [see our earlier discussion of CERW in Sect. 2.5 and Thogmartin et al.'s (2004b) reference to CERW use of canopy gaps, p. 1767].

Although the objective was arguably a *regional* model of relative abundance, combining all route stops meant that even a single detection at any one of the 50 stops would classify the entire surrounding 800-ha minimum sampling area as "occurrence" data for the associated environmental variables. Furthermore, the fact that CERW is relatively uncommon (i.e., even the most populous route had only 9 [op. cit.]) meant that even on routes where warblers were recorded, most individual stops had no birds. As a result, most of the landscape-scale (i.e., transect-wide) GIS "occurrence" data included in the relative abundance model were associated with BBS stops at which no birds were detected (see Sect. 2.4 herein and Huston 2002). In a subsequent test of CERW Atlas data, the authors found that the model "*overpredicted the occurrence of CERW at the high end of the predicted abundance and underpredicted at the low end of predicted abundance*" (41 % of CERW detections were at locations predicted to have no warblers).

If the coarseness of the Landsat imagery resulted in little heterogeneity in the habitat data for an entire route, then the inability to separate used and unused areas may be related to the resolution of the imagery. This, of course, assumes that landscape components related to habitat selection by CERW can be discerned at some higher resolution of remotely sensed imagery. However, if there were measurable (and meaningful) heterogeneity within the route sample (800 ha) at the Landsat resolution, then combining all GIS habitat information for the entire route means that the values of habitat data associated with all the unoccupied stops would tend to obscure habitat correlations with the species' presence at those few stops where they actually were detected. As Young and Hutto (2002: 110) noted, in any situation where a transect runs through a series of different cover types, combining data from all points on the transect "*would create meaningless sample units with respect to vegetation variables*" and would certainly reduce the predictive power of models derived from such data. McElhone et al. (2011), in a more recent analysis of BBS data for CERW, recognized this problem and examined temporal changes in land cover and fragmentation data at individual BBS stops, albeit using aerial photographs at a scale of 1:60000.

Thogmartin et al. (2004b) also noted the potential for autocorrelation in their BBS data and included a term to account for it in their initial analysis (see

Sect. 5.4 on Autocorrelation). When the original suite of environmental variables failed to account for the spatial structuring in the model, additional analyses
were performed that tested a supplemental suite of 95 *post hoc* variables in an
attempt to explain this structuring. Even after adding the best of these additional
variables, "*spatial structuring due to autocorrelation in* (bird) *counts explained
1.5 times* (the variation of) *the combined effect of all the environmental variables.*" (op. cit.: 1772). This strongly suggests to us that the minimum scale of
analysis (800 ha) simply was too geographically large for a species with a territory size of approximately 1–2 ha, an area within which higher resolution variables might be required to account for observed spatial structuring (see our
example in Chap. 6).

In contrast to the more typical approach above, Keller (1990) used 1:5000
stereoscopic aerial photography with a GRD of <0.75 m (NIIRS Level 6, Fig. 2.1)
and several GIS measures of the spatial arrangement of species-specific, local-
scale edge types to produce stepwise discriminant function analysis (SDFA)
models for two locally uncommon (10–15 % occurrence in 97 plot years) species in his study area, the black-throated blue warbler (*Setophaga caerulescens*)
and black-and-white warbler (*Mniotilta varia*). In tests of the models employing
an iterative data-splitting/model-building technique, SDFA accuracy for predictions of presence/absence averaged 96 and 83 %, respectively, for the two species.
Stepwise maximum r-square improvement multiple (MAXR) regression models
for the same species explained 92 and 60 %, respectively, of the variance in their
observed densities. Inclusion of several new covariates in a reanalysis of the dataset for these species has further improved the explanatory power of the MAXR
models (see Chap. 6).

With hindsight, it would seem more reasonable to develop models such as
the one above by (1) comparing habitat at individual stops where CERW were
detected (or alternatively with abundance per stop) with stops where they were
not, as per McElhone et al. (2011). This latter study, despite incorporating a BBS
stop-level analysis, did not find significant correlations of CERW population
changes with forest metrics. The authors attributed this to the limitations ("coarseness") of the 1:60000 imagery they employed and concluded that "*smaller, microhabitat features may be most important in affecting Cerulean Warbler breeding
habitat suitability*" (op. cit.: 699). This latter result suggests that the second element of such BBS studies should be (2) the inclusion of high-resolution imagery
to assess local (intraterritory) features of habitat, at least in the initial stage of
model building.

Even in the absence of these alternatives, if the original Landsat imagery in
Thogmartin et al. (2004b) was used and the minimum 100-m buffer employed was
applied as a radius of assumed detectability, this would result in an available GIS
sample area of about 3.1 ha at each stop. Note: Application of this approach would
be problematic due to year-to-year variations in exact stop locations (Thogmartin,
personal communication). Such a sample would have roughly 35 (30 × 30)
Landsat pixels for each GIS sample and would serve as a more reasonable minimum unit of analysis for habitat use. Larger radii habitat samples then could be

applied to look at effects of landscape context. Although not much spatial information would be measurable in the 100-m-radius sample, this limitation would not affect the variables that were originally considered since the only two GIS variables in the original model were "weighted median deciduous forest patch *size*" and "*proportion* of landscape in forested wetlands," neither of which considers spatial arrangement.

Note: The preceding series of observations are not limited to the referenced studies and are only meant to illustrate what we perceive as symptomatic limitations of an often employed low-image-resolution, general-landscape-metric approach to GIS-based species–habitat analysis. We additionally acknowledge the difficulty of developing such geographically extensive models to meet desired conservation objectives but note that more recent efforts incorporating occupancy modeling (MacKenzie et al. 2003; MacKenzie and Royle 2005) have met with greater success (cf. Collier et al. 2013).

As a final example of the difficulty of matching GIS landscape samples to survey samples, Donovan and Flather (2002) analyzed landscape structure at a scale of 1:250000 for 1,200 km^2 circular areas surrounding BBS routes, which sampled birds along a 32 km^2 transect. Thus, the landscape analyzed was 37 times larger than the area for which bird data were obtained. Although it can be argued that the intent was to characterize the influence of fragmentation of the spatially extensive surrounding landscape on forest interior species, the analysis did not find correlations with trends in individual forest interior species, but rather was only able to obtain a significant relationship between interior species and unfragmented landscapes when a *non-forest interior* species, the Northern Cardinal (*Cardinalis cardinalis*), was included in the regression. As noted in Sect. 3.2.3, this may have been at least partially due to inclusion of explanatory variables that did not adequately describe the spatial arrangement of all landscapes (see Fig. 2 in Donovan and Flather 2002). However, it also may reflect the fact that more than 95 % of the area analyzed was not associated with the bird data. Even if the influence of the surrounding landscape on the species or taxon is of interest, we suggest such an analysis should (1) include at least an additional local-scale component where the area of habitat, patch, biotope, or landscape analysis approximates the actual area of the bird survey (see Bart et al. 1995) and (2) the resolution of the imagery should more closely match the scale at which the animals are thought to use the landscape (Trani 2002; Van Horne 2002).

Although the acknowledged intended use for BBS data is that of long-term population trend analysis, these extensive datasets increasingly have been used to evaluate habitat associations across extensive geographic scales. In general, these and many similar landscape-scale studies correlate point-count bird occurrence data with landscape attributes of areas not associated with the bird data (Fig. 4.2). We suggest that it is inappropriate to build a habitat model where dependent data, at least at the smallest geographical scale, do not coincide with samples of explanatory data at a species-specific minimum analytical unit, unless the research question explicitly pertains to site context (e.g., Pearman 2002).

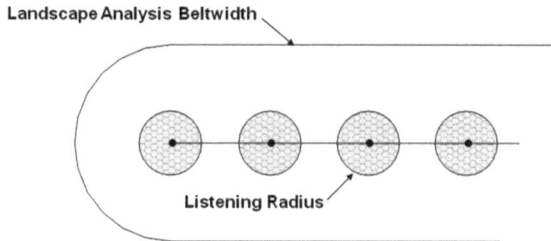

Fig. 4.2 BBS route terminus, illustrating an approximately 150-m listening radius sampled around 4 stops versus a nominal 0.90-km Landsat belt width analyzed in a GIS for habitat structure associated with the transect. Note the disparity between the area sampled for birds (*hatched*) and for habitat, which fosters development of habitat models with data not associated with the areas actually used by the species

For studies where BBS data are employed to determine habitat associations, we suggest analyses should:

1. Include buffer widths around the BBS route scaled to differences in detection across the biotopes being sampled (i.e., greater distances in more open plant communities) and when possible, use associated distance mapping (e.g., Dickson et al. 2009), and/or
2. Demonstrate statistically, at an appropriate image resolution (see below), that areas of inference are not different from areas where birds are sampled (Young and Hutto 2002).
3. Use, at least at the finest ecological scale, remotely sensed imagery with resolution of no less than NIIRS Level 5 (GRD $\leq$1.2 m) unless the area analyzed is demonstrably homogeneous from the focal taxon's perspective at this resolution (see Sect. 2.5).
4. Optimally, use the detection *radius* or a multiple of that radius up to 2 (cf. Brennan and Schnell 2007) rather than using a fixed belt width as the minimum analytical unit within which habitat structure is analyzed. Otherwise, the area for which there is no information to corroborate actual occupancy (MacKenzie et al. 2003) by the focal taxon becomes an increasingly larger percentage of the sample being used to infer such occupancy (Fig. 4.2).

4.2 Matching Metrics to Organism Location in the Landscape

We further suggest that in the absence of true organism-centered samples, researchers employing point-count approaches in GIS analyses attempt to incorporate additional, more organism-centered explanatory variables into such analyses. For example, fragmentation by woody invasion (i.e., succession) is a documented negative effect on some bird species that breed in open grasslands (e.g., Grant

(a)

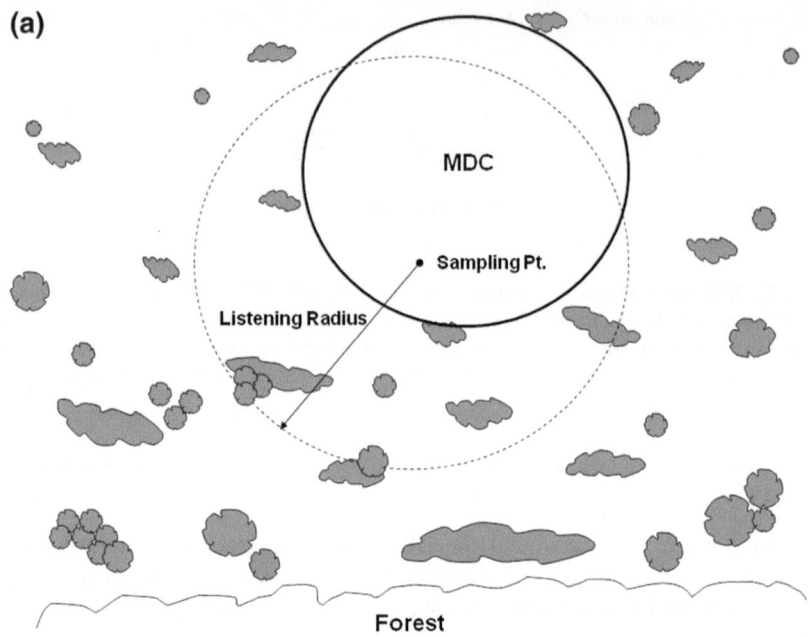

(b)

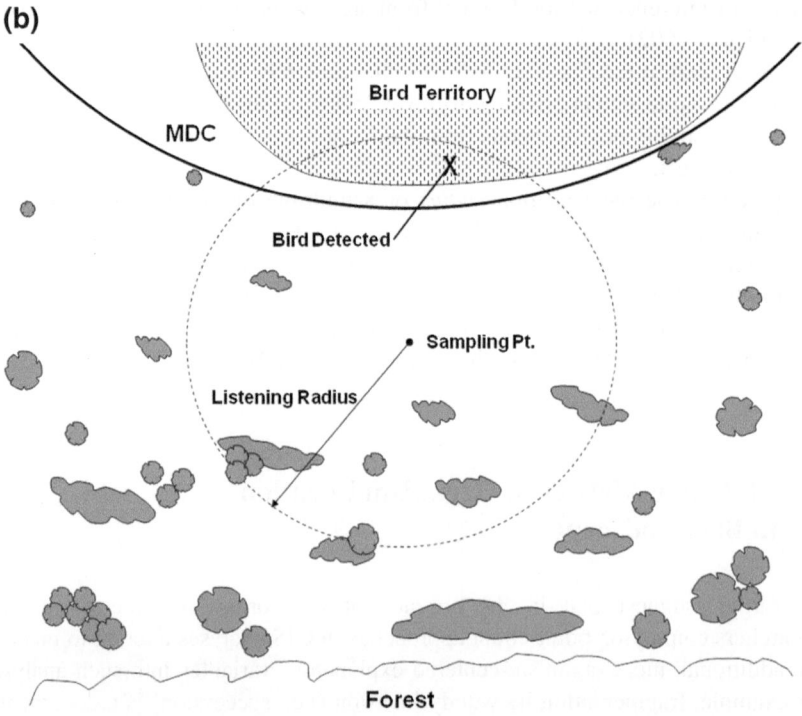

◀ **Fig. 4.3 a** Use of a spatially explicit metric, the maximum diameter circle (MDC) of open grass, in conjunction with point-count survey methods, to quantify the functional amount of open grass patch type available for grassland birds within a hypothetical early successional oldfield. For points at which no birds are recorded, the MDC must include some portion of the sampling area (i.e., listening radius). No grassland birds were detected in this sample, and the associated grassland MDC is relatively small. **b** Use of MDC to measure the functional amount of open grassland in an oldfield where a grassland bird is recorded. The MDC includes the *bird's* location for those point-count stations at which a bird is detected. Note that because both samples 4.3a and b include identical proportions and distributions of woody vegetation within the listening radius, no landscape metrics applied only *within* that radius would separate the samples. Here, it is actually a minor decrease in woody plants **outside of the point-count listening radius** that results in the presence of a portion of a territory within the radius and the subsequent detection of the species within that radius. This illustrates how the use of an additional, more spatially explicit, organism-centered variable that is not limited to calculation entirely within a fixed sampling radius avoids some of the limitations of fixed-area samples

et al. 2004), a "solid" patch type per our earlier discussion. In this setting, researchers may attempt to determine habitat preference by sampling vegetation within a fixed radius around a sampling point and statistically comparing attributes of occupied and unoccupied sampling areas (cf. Graves et al. 2010).

As depicted in Fig. 4.3, sample A contains no grassland bird, while sample B does. However, because both samples include identical proportions and distributions of woody vegetation, no metrics applied only within the listening radius would separate the samples. Sampling a larger landscape radius around the point to elucidate the habitat differences would introduce additional vegetation information predominantly unrelated to the attributes of the bird territory that resulted in the contact in Fig. 4.3b. Use of proportions or indices still would leave even the larger samples statistically inseparable.

One approach to quantify the difference between the samples would be to measure the largest circle of open grassland (i.e., MDC, see Sect. 3.2.3) that (1) includes *any portion of the sampling area* (i.e., listening radius) for those sample points at which no birds are heard (Fig. 4.3a) or (2) includes the *bird's* location for those points where a bird is heard (Fig. 4.3b). This procedure, or alternatively, a gradient analysis of the distribution of open grass from the forest to the completely open portion of the grassland, would help capture the association, if in fact there is one, between the grassland bird and more open areas (=larger MDC). Thus, use of an additional, more spatially explicit, organism-centered variable that is not limited to calculation entirely within a fixed sampling radius avoids some of the limitations of such samples. This type of approach would be applicable in forested landscapes as well, for example, as an alternative to using proportions in the 150- and 500-m sampling radii in Betts et al. (2010), or in many BBS samples.

Lastly, the type of error illustrated in Fig. 4.1b frequently obtains when vegetation samples are centered on nests or particular song posts within territories; but, such studies most often focus on attributes of those specific features (e.g., nest sites) and are less concerned with statistically describing the entire territory (e.g., Winter et al. 2005; Collins et al. 2009; Schill and Yahner 2009). Similarly, studies that focus on the influence of a site's context (e.g., Bakermans and Rodewald 2006;

Cornell and Donovan 2010; Kennedy et al. 2011; Thompson et al. 2012) deliberately examine the vegetation composition of larger surrounding areas, typically within 1–5 km^2, and do not assume that the focal taxon uses these areas, only that the organism(s) may be influenced by the area's composition.

4.3 Controlling for One or More Variables

A corollary of extrapolating species occurrence or abundance data from point samples across supposedly homogeneous GIS landscapes is the practice of making inferences from ground-based samples that larger "similar" landscapes are identical (i.e., statistically similar) structurally. This approach has been used since Forman et al. (1976) and Galli et al. (1976) to control for the effects of structural heterogeneity in tests of the effect of "island" size on virtual habitat islands (e.g., woodlots, lakes, grasslands, clear-cuts). However, the approach has rarely employed landscape metrics to analyze for spatial heterogeneity of samples and instead typically relies on ground-based measures of foliage height diversity (FHD), % cover, stem density, etc.

As we noted earlier, landscapes with identical percentages or densities of particular components can have very different spatial arrangements that can in turn produce different species associations. It is this erroneously assumed but unaccounted for internal spatial heterogeneity/patterning, and other factors such as site history (e.g., Ernoult et al. 2006), patch functional size, and productivity (Keller 1986) that, we suggest, frequently produce much of the scatter in graphs of species richness versus island size (cf. Rudnicky and Hunter 1993). In this latter example, the authors also assumed that clear-cuts ranging in age from post-cut years 3–10 were essentially identical structurally. Numerous studies have demonstrated that vegetative structure, as well as bird species richness and composition, change dramatically during this short time frame in northeastern forests (e.g., Titterington et al. 1979; DeGraaf 1991; King et al. 2001; Keller et al. 2003; Martin et al. 2007; Schlossberg and King 2009), independent of clear-cut size. It would appear that in such geographically extensive studies, great care must be exercised if researchers seek to truly control for one or more potential explanatory variables.

Chapter 5
Refining Habitat Specificity

Prior to any species–habitat investigation, the researcher chooses the focal taxon and defines the objectives of the study. At that point, the variety of habitat or landscape descriptors available to an investigator is virtually unlimited; and theoretically, to avoid statistical bias, a broad range of potential explanatory variables should be quantified to describe the full gradient of attributes (physical, chemical, and biological) of the environment in which the focal taxon is found. This is, in fact, entirely appropriate for assemblage level studies. However, including large numbers of explanatory variables invariably leads to (1) multicollinearity of some of these variables (Capen et al. 1986; Menard 1995; Graham 2003) and (2) potentially spurious correlations with dependent variables when the number of explanatory variables represents too large a fraction of the independent data set (Tabachnick and Fidell 1996; Pearman 2002). Note: This can occur even with the use of an information theoretic approach (Burnham and Anderson 2002) to systematically test all combinations of explanatory variables selected. It is still up to the researcher to select the initial set of explanatory variables to be included in the investigation (Box and Hill 1967; Burnham and Anderson 2002). Van Horne (2002) suggested this process of a priori variable selection superseded later analytic processes in importance due to the tremendous number of variables that can be generated in GIS-based analyses. We agree.

To address the question of variable selection statistically, researchers seek both suitably large independent data sets and a statistically reasonable, yet ecologically comprehensive, number of explanatory variables. Although simple in concept, addressing this latter objective is challenging, but can be one of the most intellectually engaging aspects of ecological research.

© The Author(s) 2014
J.K. Keller and C.R. Smith, *Improving GIS-based Wildlife-Habitat Analysis*,
SpringerBriefs in Ecology, DOI 10.1007/978-3-319-09608-7_5

Questions:

1. How do we increase the likelihood of including potentially useful explanatory variables? In this regard, how useful or potentially biased is a researcher's
   existing knowledge of the subject matter?
2. Are all potential solid and edge landscape components or component combinations (i.e., patch types) equally important to measure?
3. How much lumping and splitting of landscape component types is appropriate?

5.1   How Do We Increase the Likelihood of Including Potentially Useful Explanatory Variables?

Short answer—For investigations that include landscape (i.e., GIS-based) metrics, start with a sufficiently high-resolution remotely sensed image (identified per
our earlier discussions), a mapping scale interpretable from the perspective of the
focal taxon, and select explanatory variables related to known species' function
and a best estimate of the scale of habitat use.

5.1.1   Identifying a Statistically Acceptable Number of Ecologically Relevant Explanatory Variables. Can We Avoid Bias?

As noted above, inclusion of a large number of explanatory variables in the
analysis has at least two potential drawbacks. With regard to multicollinearity, researchers have used both best professional judgment (e.g., Lichstein
et al. 2002a; Pillsbury and Miller 2008; Verschuyl et al. 2008; Habibzadeh et al.
2013) and statistical formalizations of these arguments (cf. Capen et al. 1986;
Saab 1999; Graham 2003; Johnson et al. 2004) to (1) reduce multicollinearity
within a set of explanatory variables and (2) select variables that have good biological interpretability for use in resource management. However, even prior to
the use of any acknowledged procedures to reduce multicollinearity, we suggest
that, almost universally, researchers consciously or unconsciously apply a priori
subject matter knowledge to exclude descriptors that intuitively are unrelated
to the species of interest. For example, someone studying habitat preferences
of the Grasshopper Sparrow (*Ammodramus savannarum*) is unlikely to measure the distance to the nearest body of water. However, within the same landscape, a researcher studying Marbled Salamander (*Ambystoma opacum*) habitat
use would certainly consider this variable potentially important. In either case,
the investigator is not likely to consider it necessary to defend the exclusion or

inclusion (i.e., relevancy) of this variable for landscape characterization for the taxon of interest.

All of this is to emphasize that at some point, knowledge of the taxon's function and spatial use of habitat will come into play in selection of explanatory variables. While we agree that subjective elimination of variables can be risky, formal approaches can often produce similar results (Green 1979; Capen et al. 1986; Johnson et al. 2004). In some cases, particularly with the almost unlimited number of landscape metrics available, even formal attempts to select the most appropriate variables produce ambiguous results, leaving investigators to select "preferred" variables (cf. McGarigal and McComb 1995).

In other cases, investigators have eschewed any explanation of the variable selection process and simply selected a precise set of variables they identified a priori as related to the focal species' use of the landscape (e.g., Thompson and McGarigal 2002; Thogmartin et al. 2004b; Stoddard and Hayes 2005; Rehm and Baldassarre 2006; Hallworth et al. 2008). For example, among the limited set of potential predictors of eagle habitat use they selected, Thompson and McGarigal (2002) measured tree canopies but not understory vegetation. Although unexplained, this is presumably because eagles are not understory species and only *canopy* presence and distribution, particularly near open water, is important to their ecological function.

This "cut-to-the-chase" approach runs the risk of missing certain variables, at whatever scale, that may have explanatory power and/or biasing variable selection. However, it often focuses, and we think appropriately so, on a subset of descriptors related to the species' function or requirements. At the habitat (local) and patch scale, this includes quantification of individual components and combinations of components that characterize foraging sites (based on tactics), nest/den sites, prey abundance, etc. These, in turn, are related to thermoregulatory and other energetic considerations (e.g., foraging efficiency, territorial defense) that dictate habitat selection, territory shape, and use. At the landscape scale, characterization of attributes such as insularity, connectivity, regional biotope availability, and contextual composition and structure may provide additional explanation of the influence of landscape context on species, guild, or assemblage distribution and abundance (Lee et al. 2002; MacFaden and Capen 2002; Bakermans and Rodewald 2006; Kays et al. 2008; Cornell and Donovan 2010; Kennedy et al. 2011).

Ultimately, the effect of researcher bias in variable selection will be judged by the utility of the resulting models in resource management. If demonstrably unbiased attempts to model species–habitat relationships result in models with little explanatory power that fail to enhance conservation efforts, then what has been gained by being unbiased? We suggest efforts to refine variable selection based on best available understanding of species natural history, particularly function and scale of habitat selection, outweigh potential statistical bias in such processes. This is not to imply that models must be biased to have good explanatory power.

5.1.2   Matching Both Spatial and Temporal Scales of Explanatory Variables to the Organizational Level of the Analysis

When trying to assess biodiversity at the landscape scale using a broad range of variables to quantify all aspects of that landscape and relating those attributes to an entire assemblage, one invariably includes variables that are weakly or completely uncorrelated with a given species in the assemblage. For example, a habitat variable such as litter depth may have little or no correlation with the presence and abundance of canopy species of birds but may be quite important to forest floor species (e.g., Ovenbird, forest floor mammals, and ground beetles). Similarly, in GIS analyses, extent of a particular plant community such as forest may have little to do with presence and abundance of species not associated with forest interior (cf. Table 5 species list in Cushman and McGarigal 2003; and discussion of early successional vs. late-successional data splitting on p. 234 of Hagan and Meehan 2002).

Secondly, measuring area or proportion, even of a plant community known to be associated with species of interest, without considering optimal spatial arrangement (energetics) still may result in poor species–community correlations (cf. forest interior species in Table 3 of Penhollow and Stauffer 2000; Betts et al. 2006). This suggests the advisability of conducting analyses at the level of individual species or guilds, which allows more precise matching of dependent and explanatory variables (cf. Lee et al. 2002), rather than at the level of more comprehensive assemblages. A discussion of the relative merits of analysis at the level of individual species, guilds, or assemblages is included in Appendix B.

As noted earlier, landscape-level descriptors also may correlate with a species occurrence over only a small range of the variable's distribution. For example, changes in species composition during early succession are often very rapid (e.g., Conner and Adkisson 1975; DeGraaf 1991; Keller et al. 2003), and optimal habitat conditions for a particular species may exist for only a few years (e.g., Confer and Knapp 1979; Probst and Weinrich 1993; Bollinger 1995; Martin et al. 2007; Schlossberg and King 2009; Anich et al. 2011). Thus, more precise variables that describe the architecture and spatial distribution of habitat components at specific points during succession, such as the vertical distribution and density of leaf area, and presence and density of particular types of edge (e.g., deciduous trees and deciduous shrubs adjacent to grass) are needed to better identify management prescriptions and appropriate disturbance intervals to maintain habitat for the species or guilds of interest in ephemeral plant communities (Williamson 1970; Keller 1986; Probst 1988; Bollinger 1995; DeGraaf and Miller 1996; King et al. 2001; Dechant et al. 2003; Keller et al. 2003; Martin et al. 2007; Donner et al. 2008; Hallworth et al. 2008; Schlossberg and King 2009). More general variables such as percent forest cover, or even percent cover of a particular forest age class (e.g., sapling-pole stand) are likely insufficient for such purposes. For example, see Probst's (1988) discussion of the inadequacy of tree density in determining habitat availability for Kirtland's Warbler (*Setophaga kirtlandii*) due to variability in tree *spacing* (e.g., Fig. 3.1 herein).

The need for fine tuning measures of structure, particularly during the rapid vegetation changes associated with early succession is illustrated by Martin et al. (2007), who studied GWWA use of aspen clear-cuts and several other early successional or shrub-stage plant communities in Wisconsin. They initially grouped all the early stage (3- to 10-year-old) clear-cuts into a *single*-stand-type and aggregated explanatory variables across these variously aged plots based on *average* stand structure. However, when they examined year-to-year changes on individual plots, they noted succession-related trends (i.e., initial increases followed by decreases) in density of GWWA as regenerating stand suitability quickly waxed and then waned. Such a time series approach allows assessment of (1) the successional age and vegetation configuration at which the clear-cuts provide optimal habitat for GWWA and (2) how long after cutting the stands remain viable as GWWA habitat. This latter information provides a basis for stand management to maintain metapopulations (cf. Probst 1986; Schlossberg and King 2009).

Similarly, there may be internal heterogeneity within grasslands that is critical to species distributions within these communities (Wiens 1974), but which is difficult to quantify at a typically lower Landsat image resolution. Obtaining better correlations of species use patterns within a biotope may require the use of higher resolution imagery, or techniques such as spectral (textural) analysis (e.g., Guo et al. 2004; Bellis et al. 2008) or LiDAR (Vierling et al. 2008; Goetz et al. 2010) that provide an alternate means of quantifying internal heterogeneity. Recall also that even if higher resolution imagery is viewed by the researcher at too small a mapping scale (i.e., not at the scale viewed by the organism), it still may be difficult for the researcher to discern this heterogeneity and thus to appreciate its potential influence on the species. Therefore, both image resolution (GRD) and the ability of the researcher to observe it (mapping scale) are important to consider.

Finally, many studies of species–habitat relationships are still conducted solely at the local level using only ground-based variables. This is often entirely appropriate based on the research question (e.g., Collins et al. 2009). However, in studies of game species or those of conservation interest where the goal is population management, the inclusion of analysis of habitat spatial arrangement may usefully refine management prescriptions (op. cit., Stoddard and Hayes 2005).

5.2   Are All Solid and Edge Landscape Components or Component Combinations (i.e., Patch Types) Equally Important to Measure?

Short answer—No. Measure those that most clearly relate to species' or assemblage function.

As you read this, look out a window in your office at the landscape outside. How many different (solid) landscape elements can you identify? A mature deciduous tree? More discretely, its canopy, trunk and the height distribution of its leaves? The understory beneath it? A coniferous tree? Subcanopy tree? A low shrub? A tall shrub? Are the shrubs in the open or in the understory of the trees? Nectar producing flowers? Mowed lawn? Sidewalk? A mulched bed? Bare ground? A water feature? Adjacent buildings? Their rooftops? Air? Now, additionally consider the edges between them. Various combinations of these elements (and others) and their spatial arrangement represent important habitat structure to various species due to their relevance to the species' function, energetic efficiency, and reproductive success. Simultaneously, other elements in the same setting have little to no influence on habitat use by a particular species (Keller and Smith 1983). Additionally, if this same landscape is viewed at a more coarse-grained (i.e., lower resolution) scale as campus, suburban yard, city block or park, the elements critical to a particular species' occurrence are often unidentifiable, obscured within a landscape classified as a single type, attributes of which may correlate poorly with the presence and abundance of the focal taxon (see Chap. 2 discussion of image resolution and analytical scales).

5.2.1 The Case of Edge Versus Ecotone

We have repeatedly noted the potential weaknesses of using only variables that broadly characterize an entire plant community or biotope to develop models for individual species or higher organizational levels of interest. In developing explanatory variables, consequences of the decision to (1) use the broad characterization approach or (2) dissect biotopes into their component parts and reassemble those components to describe various "place" entities (e.g., habitats and patches) are probably no better illustrated than in the case of how best to interpret and measure edges. As discussed previously, we find the constricted usage of edge in the sense of an ecotone limiting conceptually, analytically, and ultimately, from a management perspective. We further argue that limiting application of the term only to an ecotone scale leads to a loss of information and has resulted in confusion regarding species–habitat relationships.

Once again, as you are reading this, look at all the edge types on the desk in front of you and in the room around you. Though perhaps mostly abiotic, some are likely important to particular species or taxa. In the case of such an indoor setting, these are likely insects, spiders, or silverfish (at least in our offices). However, many of the edges you can see are less useful and would be expected to have little explanatory power, even for the species in your office. Likewise, there are both meaningful and relatively meaningless edges for particular species or assemblages in exterior landscapes (cf. Fig. 1.2).

Recognizing the variety of edges that permeate a landscape, it seems reasonable that to investigate edge influences on species, researchers should first identify a suite of potentially germane edge types (i.e., related to the taxon's function at a scale commensurate with the taxon's use of the environment). These interfaces between various landscape components (trees, shrubs, grass, rocks, water, etc.) can then be tested for their explanatory power on species such as Blue-winged Warbler (*Vermivora pinus*), Willow Flycatcher (*Empidonax trailii*), and Pipefish (*Stigmatopora argus*) (Macreadie et al. 2010) associated with habitats composed of edges.

As one example of the interpretational consequences of viewing edge only in the ecotone sense, Schlossberg and King (2008) analyzed data for 17 species of shrubland nesting birds on regenerating clear-cuts, demonstrated that these species avoided "edges," and concluded they were thus not edge species. However, the analysis identified "edge" in the above-referenced limited sense of an ecotone, in this case the boundary between a clear-cut and the surrounding forest (i.e., canopy trees adjacent to dense shrubs). This narrow definition fails to consider that this is just one of *many* types of edge that occur in such landscapes. Early successional clear-cuts also contain many *internal* edge types (as well as solid patch types such as the dense shrubs mentioned earlier as used by the common yellowthroat) that explain the tendency of certain species to avoid the clear-cut perimeter and instead hold entirely or predominantly internal territories. For example, Keller (1986, 1990) identified more than 40 different early succession-associated edge types, various combinations of which were highly correlated with the richness and density of eight different avian guilds of edge species on young clear-cuts and oldfields, several of which were referenced in the preceding chapter. In Schlossberg and King's (2008) study, the assumption that the landscape contained only one major type of edge to which birds responded led them to a totally different conclusion than if the landscape were viewed as including many edge types, both within and between plant communities.

Similarly, Kasumovic et al. (2009) analyzed Least Flycatcher (*Empidonax minimus*) breeding densities and reproductive success in what they termed "fragmented" (i.e., edge dominated) versus "continuous" (i.e., solid) forest. They concluded that "*habitat fragmentation does not appear to affect realized reproductive success of male*" flycatchers. However, examination of their definitions of what constituted fragmented versus continuous reinforces our earlier discussions of scale and edge specificity. Figure 1 of the study reveals that the "edge" used to define the fragmented forest condition is the *ecotone* between forested areas and other adjacent land covers (Fig. 1a), in this case a lake and cleared areas of development. However, the forest area characterized as "continuous" (Fig. 1b) actually is even more fragmented than the "fragmented" landscape; in this case, internally by a road network and more than 30 cleared lots, which serve to form numerous canopy/subcanopy gaps. Thus, we do not find it surprising that observed flycatcher breeding densities were more than 50 % higher in this so-called continuous landscape than in the ecotone-defined "fragmented" landscape. Rather, we suggest this is because, measured at the scale of use by the flycatcher, territories located in the "continuous" landscape have a much higher density of the subcanopy-opening edges used by this sallying-forager (Sherry 1979). At this species-oriented scale, the portion of the "continuous" forest where the flycatcher territories are located is not really continuous at all. These higher observed densities are also completely consistent with the concept that flycatchers, as a group, are edge species and not ones we would expect to be most abundant or successful in more closed-canopy (or closed subcanopy) forests, as is implied by the term "continuous" (e.g., see absence of flycatchers in "mature" forests in Table 2 of King and DeGraaf 2000). In this study, the internal (within-biotope) edge present within the forest (i.e., solid patch) simply was not recognized as edge at the chosen scale of analysis.

Similarly, see the negative correlation of edge, again measured as ecotone on Landsat imagery, with least flycatcher occurrence in Fig. 4 of Villard et al. (1999). Lest these examples appear to be simply semantics, consider the consequences of relying on studies with potential conceptual confusion to develop management prescriptions for species or guilds of interest (Strayer et al. 2003). In this case, managing for closed-canopy/subcanopy forest to promote high densities of least flycatchers would likely be a failed strategy (see Degraaf et al. 1998; King and DeGraaf 2000; Holmes and Sherry 2001).

In addition, the makeup of the edges associated with high densities of Least Flycatcher in Kasumovic et al. (2009) and King and DeGraaf (2000) illustrates another point about edge specificity. In the former study, higher densities of the flycatcher are associated with canopy openings in a developed area with little or no shrub layer within those openings, only lawn, house or road. In the latter study of shelterwoods, the shrub layer in the forest openings is densely stocked with regenerating sprouts and root suckers. Thus, what is on the ground in the "opening" portion of the edge is mostly irrelevant. The most important portion of the edge for the least flycatcher would appear to be the canopy/subcanopy adjacent to *open air* (i.e., the edge type at the height used by the bird). This habitat element and its high density per unit area are the common attributes of the highest-density locations of the flycatcher in both studies.

Finally, broadly defined edge types can be useful when they represent a functionally important type at a scale meaningful to the focal taxon. Wilson and Watts (2008) found higher densities of Whip-poor-will (*Antrostomus vociferus*) at the edge between regenerating pine stands and adjacent older forested stands in North Carolina than in either the forested areas alone or along edges between older stands. They concluded that the regenerating stands offered increased foraging opportunities while the adjacent forests offered better roosting and nesting habitat. In this case, identifying only major adjacent biotopes, represented by two disparate stand age classes, and the ecotones between them sufficiently characterized the within-territory habitat components used by this larger (than in the previous study), more wide-ranging species. Subdividing regeneration areas into finer components such as individual pine trees, herbaceous openings, and shrubs may not have improved the results or interpretations, and likely would not have changed the management recommendations.

5.3 How Much Lumping or Splitting of Types Is Appropriate?

As can be seen in the Wilson and Watts (2008) Whip-poor-will study, it is not always necessary to subdivide plant community (biotope) types and ages in GIS analyses. However, failure to subdivide community types *finely enough*, often due to the use of lower resolution imagery, can reduce the explanatory power of models, whether of species richness or individual species–habitat associations (Orrock et al. 2000; Thompson and McGarigal 2002; Trani 2002).

For example, Chandler et al. (2009), examined use of beaver meadows by nine species of shrub-scrub birds. Among the variables, they measured were the size and shape of the meadows. Although a beaver meadow can be interpreted as a single community, it also, especially considering space use by passerine birds, could logically be interpreted as a series of communities defined by the underlying hydrologic gradient. This latter interpretation may, in fact, foster development of more specific variables that improve the analysis. In Chandler et al.'s study, rather than measuring the size and shape of solid or edge patch types representing more species-specific subsets of components within the meadows, the authors measured size and shape of the *entire* meadow, thus aggregating (equating), for size and shape, all the internal structural types (i.e., open water, emergent marsh, shrub-scrub, and early successional forest) that fall under the broad heading of beaver meadow. In the resulting analysis, only three of the nine species studied showed even a weak relationship to meadow size and only two were correlated with shape. This type of habitat aggregation into *community-level* variables, which are then applied back to the level of *individual species* is quite common (see next example) and as suggested by the examples in the previous chapters, likely reduces the explanatory power of the models generated. For example, the Eastern Kingbird (*Tyrannus tyrannus*), a species associated with edges more fine-grained than those quantified, had the poorest habitat model ($R^2 = 0.08$). Overall, of the single best models for each species examined, only 2 of 9 explained more than 50 % of the variance of a given species' abundance. In our opinion, this underutilizes the potential of carefully gathered, extensive data sets such as this one.

Lapin et al.'s (2013) study of Connecticut Warbler (COWA) habitat relationships at three different scales appears to be a similar example of how insufficiently resolved imagery limits classification and quantification of within community heterogeneity, in this case that of a spruce bog. Here, the referenced habitat description (op. cit.: 169) suggests the COWA is actually an *edge* species, one associated with shrubs and grass adjacent to spruce trees within the bogs it "primarily" inhabits. However, due to the 30-m image resolution and even larger mmu's employed (cf. Fig. 4 therein), (1) analysis of potentially important species-specific edge associations (e.g., using DEAC) is precluded; and (2) the bird is instead inferentially classified as a *solid* patch species due to its association with "lowland coniferous forest," one of 13 broad cover types classified in the GIS. The models produced at this L-resolution are characterized by (1) modest predictive power (average $R^2 = 0.23$; $R^2 < 0.37$ for 5 of 6 models); (2) several strong negative associations of COWA with unused cover types (e.g., upland deciduous forest), which provide little management guidance; and (3) the misleadingly *negative* association of edge (measured as total ecotone-scale edge, op. cit. Table 1) with a bird that, based on numerous literature accounts, appears to be a more fine-grained edge-species. Like the preceding example, this is a well-intended study of a very useful data set which, in our opinion, has additional potential for guiding management of this designated "stewardship species."

As a final example, Betts et al. (2006) studied habitat associations of 21 species of birds in forest successional seres in eastern Canada. Although explanatory variables were identified at three scales (local on the ground, local air photo, and landscape air photo), many of the variables were either the type of generalized plant community descriptors discussed previously or were proportions. None of the air photo-derived variables considered spatial arrangement at the local scale of bird territories. Rather, only the areal extent of particular forest types within the surrounding 78 ha area (i.e., within a 500 m radius) or larger was calculated. Edge density also was measured only at these more extensive geographic scales as ecotones. In addition, landscape components, in this case edge types, or continuous data that might have provided insight to species–habitat associations were aggregated (i.e., all edges viewed as equivalent, or continuous variables reduced to categorical data), as in several of the studies above.

Although using a larger mapping scale (1:12500), and inferentially, a higher resolution image (none was specified), than many analyses of passerine assemblages incorporating GIS-based variables, only one (1) model for the 21 species studied explained more than half ($R^2 > 0.5$) of the variance associated with the observed species distributions; and 14 of the 21 models explained less than one-third ($R^2 < 0.33$) of the variance. In particular, landscape variables at best explained less than 10 % ($R^2 = 0.098$) of the variance in any of the 21 species' models. As with the preceding study, we suggest these results could be improved with higher resolution imagery and, in this case, identification of more precise edge types and inclusion of variables that better assess plant community size and shape at the local scale used by the birds under study. In contrast, Keller (1986, 1990), using high-resolution aerial photography (GRD <0.75 m, NIIRS Level 6) with stereoscopic landscape component classification as the basis of a GIS analysis, obtained average MAXR regression values of 0.721 for the species richness of 19 avian guilds he studied.

5.4 Spatial Autocorrelation

Lastly, spatial heterogeneity in a response variable is caused either by spatial dependence or spatial autocorrelation. Spatial dependence is the species response to environmental conditions such as vegetation structure that are themselves spatially structured by additional physical processes (e.g., rainfall, slope gradient and aspect, and soil type) (Wagner and Fortin 2005). These are the habitat relationships that investigators typically seek to quantify when developing species–habitat models in order to better manage species of interest. The strength of these models can be compromised by the second cause of spatial heterogeneity, spatial autocorrelation. Spatial autocorrelation reflects biotic processes such as site fidelity, social facilitation, and dispersal capabilities that produce patchiness in organism distributions and dependence in adjacent or nearby values of the dependent variable (Wagner and Fortin 2005).

This latter concept is an area of discussion in habitat studies that has attracted substantial interest (Sokal and Oden 1978; Legendre 1993; Legendre et al. 2002;

Lichstein et al. 2002b; Diniz-Filho et al. 2003; Wagner and Fortin 2005; Betts et al. 2006; Diniz-Filho et al. 2007; De Knegt et al. 2010). Autoregressive models applied to spatially autocorrelated explanatory and/or response data attempt to reduce spurious correlations of species abundance with environmental variables and account for contagion in species distributions not accounted for by habitat variables. However, as noted by Diniz-Filho et al. (2003: 55), *"If different environmental factors act at different spatial scales (see Willis and Whittaker 2002), the inclusion of the relevant environmental factors acting at each scale in the regression model should be sufficient to completely remove autocorrelation from the residuals at all scales."* De Knegt et al. (2010: 2455) explicitly analyzed the effects of scale on spatial autocorrelation and found that *"... a mismatch between the scale of analysis and the scale of a species response to its environment leads to a decrease in the portion of variation explained by environmental predictors. Moreover, it results in residual spatial autocorrelation and biased regression coefficients. This bias stems from error-predictor dependencies due to scale-mismatch, the magnitude of which depends on the interaction between the scale of landscape heterogeneity and the scale of a species' response to this heterogeneity."* The authors concluded that more important than examining for spatial autocorrelation per se was the need to examine the appropriateness of the spatial scales used in the analysis, a point we have made repeatedly [cf. our discussion of Wiens (1989a) "domains of scale" in Sect. 2.5.1].

Thus, although the amount of variance accounted for by spatially autocorrelated data varies depending on both the scale(s) of the investigation and the distribution of sample points in a given study, and autocorrelation should be routinely investigated, we concur with the interpretation of Diniz-Filho et al. (2003, 2007) and De Knegt et al. (2010) that this variance, as a proportion of total explained variance (i.e., that due to habitat, patch, and landscape variables), is likely to be much smaller than it appears if some of the problems with habitat measurement and image resolution previously described are addressed (cf. Moore et al. 2010). As noted by Thogmartin et al. (2004b: 1770) in describing the need for inclusion of autoregressive techniques, *"Ideally, spatial structuring in the model would be unnecessary, given the inclusion of a proper set of environmental covariates defining the spatial relatedness between counts."*

As one example, in their comprehensive study of three species of small passerines, Lichstein et al. (2002b) (1) sampled local vegetation within 10 m of point count listening stations, (2) examined the influence of a number of landscape variables within a 250 m radius of the sampling point and (3) examined the possible effect of spatial autocorrelation among nearby sampling points. While we readily acknowledge the potential autocorrelative effects produced by site fidelity, social facilitation, and dispersal capabilities (Harvey et al. 1979; Bedard and LaPointe 1984; Sikes and Arnold 1984; Keller 1990; Ahlering and Faaborg 2006; Campomizzi et al. 2008) and the potential for these behaviors to reduce the observed strength of species–habitat relationships, we argue that: (1) Due to the relatively low resolution of the photos in this analysis (1:24000 mapping scale), some landscape features critical to the small passerines studied were likely not measurable. (2) The potential for including misclassified vegetation samples (i.e., from areas outside of the territory of a detected species; see Chap. 4) at the local scale reduces the strength of the observed species–habitat correlations. (3) Both the local- and landscape-scale explanatory variables chosen are not explicit enough to further reduce the amount of unexplained variance in the regression models (see Chap. 3). (4) Inclusion of a measure of site insularity (i.e., distance from similar habitat) might have accounted for a portion of the behavioral influence on spatial autocorrelation observed (see Keller 1990 and Chap. 6 herein). In our opinion, each of these factors contributes to some of the inferred influences of autocorrelation (i.e., unmeasured habitat variables) found in this study and in many such studies (cf., Cressie 1993, reviewed by Campomizzi et al. 2008).

Chapter 6
An Example Using High-Resolution Imagery and Taxon-Specific Variables

Species-level analyses

As an example of the high-resolution, taxon-specific, spatially explicit variable approach we have suggested, we compare the explanatory power of four commonly used landscape metrics with a set of more taxon-specific covariates in describing local- and patch scale habitat relationships of seven species of breeding birds, all of which also were studied either by Lichstein et al. (2002a, b), Betts et al. (2006), and/or Howell et al. (2008). Specifically, what attributes of landscape composition, structure, and spatial arrangement dictate the occurrence and density of each of the seven species within the landscape mosaic of a 4,500 ha state game management area in central New York? How different are the results (explanatory power, biological interpretability, inferences) between more species-specific, spatially more explicit variables and the types of landscape metrics more commonly in use at present? Additionally, which results provide the most useful information on management procedures to promote these species, several of which are exhibiting population declines at regional and/or range-wide scales?

Bird data are spot-map estimates of breeding density and presence collected for an average of 4 years on 23 plots (total of 97 plot years) ranging in age from young (2 year) clear-cuts and oldfields to mature northern hardwood–hemlock forests in central New York (Keller et al. 2003). Bird and vegetation sampling are described in Keller et al. (2003). All standardized bird abbreviations referenced in the following text are listed in Table 6.1. Imagery acquisition, GIS software, and analysis are described in Keller (1990). The imagery (e.g., Fig. 2.1), which had resolution (GRD) to <0.75 m (NIIRS Level 6), allowed identification of individual small shrubs and saplings. Grid cells, each representing 100 m^2 on the ground (Fig. 2.1 hexagonal cells), were classified to one of 16 landscape component types (Appendix C) on 1:2000 georectified base maps using 1:5000 black

© The Author(s) 2014
J.K. Keller and C.R. Smith, *Improving GIS-based Wildlife-Habitat Analysis*,
SpringerBriefs in Ecology, DOI 10.1007/978-3-319-09608-7_6

and white stereoscopic aerial photography. The classified 1:2000 imagery formed the basis for the GIS analysis.

As we have discussed, most species–habitat studies employing GIS, whether at local, landscape, or range-wide geographic scales, (1) use lower resolution imagery, (2) include landscape metrics that typically do not consider actual spatial arrangement, and (3) consider only ecotone scale edge attributes. Identified edge types are then generally aggregated (i.e., all edge is equal) or, at best, described by contrasts between adjacent landscape components, which results in an a priori loss of information. The current analysis differs in two principle ways from most of those discussed. First, our use of high-resolution, stereoscopic aerial photography allows landscape components to be more finely subdivided (Appendix C), which in turn allows quantification of more precisely defined solid and edge patch types (i.e., combinations of landscape components) related to observed species' use of the landscape (Appendix D). Exploratory patch types, solid or edge, each previously identified as associated with the presence of one or more of the seven bird species considered here (Keller 1986, 1990; Keller et al. 2003), are listed in Table 6.1.

Table 6.1 Exploratory GIS-based patch types associated with seven species of breeding birds at the Connecticut Hill WMA in central New York

Patch Type Number[a]	Description	LA Profile[b]	Species[c] Principal associates	Secondary associates
1	Deciduous dense shrubs	Low	CSWA	
4	Open shrub (shrub mixed with grass)	Low	HOWR	INBU, NAWA
9	Shrub/Opening	Low	HOWR, INBU, NAWA	
10	Sapling/Opening	Low	ALFL	HOWR, INBU, NAWA
12	Canopy/Shrub	Low		BAWW
13	Northern Hardwood-Hemlock/Shrub	Low	BAWW, BTBW	
14	Shrub-sapling/Opening	Low	INBU	HOWR, NAWA

See Appendix C for the landscape component classification system and Appendix D for the combination of individual components composing each patch type. Projected associations are based on a literature review, observations during 8 field seasons, and results in Keller (1986)

[a] For reference, the patch numbering system follows Keller (1986, 1990). Patches 1 and 4 are solid types (see Working Definitions). Patches 9, 10, 12, 13, and 14 (i.e., those whose description includes a "/") are edge types

[b] All patch types in this example used leaf area estimates from the 0–3 m (Low) portion of the vertical foliage profile to calculate the variables DEACLA, DEACLAratio, EIDCLA, EIDCLAratio, MDCLA, and MDCLAratio (see Table 6.2)

[c] ALFL = Alder Flycatcher *Empidonax alnorum*; BAWW = Black and White Warbler *Mniotilta varia*; BTBW = Black-throated Blue Warbler *Setophaga caerulescens*; CSWA = Chestnut-sided Warbler *Setophaga pensylvanica*; HOWR = House Wren *Troglodytes aedon*; INBU = Indigo Bunting *Passerina cyanea*; NAWA = Nashville Warbler *Oreothlypis ruficapilla*

Because all landscape metrics in this comparison are calculated from the same GIS data set derived from high-resolution aerial photography (Keller 1990), we are not formally testing the influence of image resolution. This brings us to the second principle difference in this analysis—variable selection. Table 6.2 lists a set of ground-based (e.g., leaf area) and/or GIS-based variables [e.g., maximum diameter circle (MDC) for dense low shrubs = MDC1] that characterize aspects of the composition, architecture, and spatial arrangement of each of the exploratory patch types identified in Table 6.1. Many of these variables are novel in comparison to the studies we have reviewed, and collectively, they address the many short-comings of landscape metrics based on proportions, density, and/or indices, which we enumerated in Chaps. 3–5.

Simple Correlations

We initially correlated four general plot- or patch-scale landscape metrics with the density and presence of the seven bird species (Table 6.3). Plot size is self explanatory. Plot Edge Density and the Plot Edge Index (see FRAGSTATS equivalents in Table 6.3) consider all of the edges that occur on the plot [i.e., both internal and the plot perimeter when the perimeter represents a different landscape component type (e.g., an adjacent plant community)]. Plot shape uses only the plot perimeter in its calculation. Of the 56 simple linear correlations (r), 28 each with density and presence shown in Table 6.3, 26 (46 %) are significant. Sixteen (29 %) of these are highly significant ($p < 0.01$).

By comparison, Tables 6.4 and 6.5 illustrate simple linear correlations (r) of a range of more species-specific variables with the density and presence, respectively, of the seven species. Note that for all seven species, the highest correlations with the more species-specific explanatory variables (Tables 6.4 and 6.5) are substantially higher than those with the general landscape descriptors (Table 6.3) for the corresponding response variables. For example, the highest correlation of a general landscape metric with either density or species presence is the correlation of Plot Edge Density with the density of BTBW ($r = -0.54$), a highly significant correlation. This explanatory variable calculates the density on the plot of *all edges combined*, a common practice in GIS analyses. However, compare the strength of this correlation with that of the similar Plot Edge Index (PLOTEI, Table 6.4) for the more species-specific Patch Type 13, which is northern hardwoods–hemlock/shrub edge (see Table 6.1), a subset of all edges on and adjacent to a given plot. The latter correlation with BTBW density is not only substantially higher ($r = 0.91$), but points to a specific management prescription to create habitat for the warbler.

Additionally, the correlation of the general edge density metric with BTBW density is negative, as is a similarly high correlation with BTBW presence, implying that the bird is negatively associated with increasing amounts of edge in general. Yet, all of the correlations for the BTBW with more species-specific edge measures in Tables 6.4 and 6.5 are positive and support the classification of the species as an edge bird associated with the clustered distribution of a specific type of edge, northern hardwoods–hemlock canopy adjacent to dense sprouts or shrubs, rather than negatively with the density of all edge.

Table 6.2 Definitions of variables used to illustrate the application of species-scaled GIS and ground-based explanatory variables to development of species–habitat models

Variable	Description
DEAC (8–16)[a]	Diameter (m) of the equivalent area circle in an "edge" patch type (T). See text and Fig. 3.5
DEAC (8–16) ÷ PS	DEAC ÷ size of the plot (=PS) on which that DEAC occurred. See text
DEACLA (8–16)	DEAC × LA (cm^2/m^2) in a specified "edge" patch type. Table 6.1
DEACLA (8–16) ÷ PS	DEACLA ÷ size of the plot on which that DEAC occurred. See text
DEACLALowMid (8–16)	DEAC in a specified "edge" patch type × LA from 0 to 7 m
DEACLAratio (8–16)	DEAC × LAratio for a specified "edge" patch type. See LAratio definition
DEACLAratio (8–16) ÷ PS	DEACLAratio ÷ size of the plot on which that DEAC occurred. See text
DIST (1–16)	Distance (m) from the plot to the nearest area where the species in question was observed holding a territory
EI	Edge index = # edges of patch (Type T) within the DEAC selected by the program ESCAN ÷ (area sampled)$^{1/2}$. See Figs. 3.4 and 3.5, and Table 3.1
EIDCLA (8–16)	Edge index (EI) × DEAC (DC) × LA in a specified "edge" patch type. See Table 6.1
EIDCLA (8–16) ÷ PS	EIDCLA ÷ size of the plot on which that DEAC occurred. See text
EIDCLAratio (8–16)	EI × DEAC × LAratio for a specified "edge" patch type
EIDCLAratio ÷ PS	EIDCLAratio ÷ size of the plot on which that DEAC occurred. See text
EIDEAC (8–16)	EI × DEAC in an "edge" patch type. See Table 6.1
LA1M	Leaf area (cm^2/m^2) from 0 to 1 m[b]
LA1Mratio	LAratio for the 0–1 m height interval
LALOW	Leaf area (cm^2/m^2) from 0 to 3 m[b]
LALOWratio	LAratio for the 0–3 m height interval
LALOWMID	Leaf area (cm^2/m^2) from 0 to 7 m[b]
LALOWMIDratio	LAratio for the 0–7 m height interval
LAMID	Leaf area (cm^2/m^2) from 3 to 7 m[b]
LAMIDratio	LAratio for the 3–7 m height interval
LAratio	1—([Absolute value of (LA observed in a specified vertical interval on a given plot—median LA for that interval over the study area)] ÷ median LA for the interval over the study area). Assumes that an intermediate level of LA within the height interval is optimal. Values range from 0 to 1
LAHIGH	Leaf area (cm^2/m^2) >7 m[b]
MDC (1–7)	Maximum diameter (m) circle in a "solid" patch type (T). See text and Fig. 3.2
MDC ÷ PS (1–7)	MDC ÷ size of the plot on which that MDC occurred. See text
MDCLA (1–7)	MDC × LA in a specified "solid" patch type. See Table 6.1

(continued)

Table 6.2 (continued)

Variable	Description
MDCLA ÷ PS (1–7)	MDCLA ÷ size of the plot on which that MDC occurred. See text
MDCLAMID (1–7)	MDC in a specified "solid" patch type × leaf area from 3 to 7 m
MDCLAratio (1–7)	MDC × LAratio for a specified "solid" patch type
MDCLAratio ÷ PS	MDCLAratio ÷ size of the plot on which that MDC occurred. See text
NUMHAB (1–16)	Number of landscape component types actually present on a plot within a given patch type (T). See Appendix C
PLOTEI (8–16)	Plot Edge Index = # edges of patch type (T) on the plot ÷ (plot area)$^{1/2}$
VMRI (1–16)	Variance/Mean Ratio × inverse of Run's Index[c]

[a] Numbers in parentheses refer to exploratory patch types (T) (Table 6.1, Appendix D). Not all of the original 16 identified patch types (Keller 1986, 1990) are included in this example. The numbering system was maintained for reference
[b] See Keller et al. 2003
[c] See Keller 1990

Table 6.3 Simple linear correlations (r) between density (pr/40 ha) or presence of seven bird species and general landscape metrics applied at the plot level

Variable[b]	Species[a]						
	ALFL	HOWR	NAWA	CSWA	BTBW	BAWW	INBU
Density							
Plot Size	0.09	−0.12	−0.07	−0.34***	−0.05	−0.05	−0.39***
Plot Shape	−0.22	−0.05	0.00	0.03	0.15	0.15	0.03
Plot Edge Density	0.23*	0.15	0.24*	−0.40***	−0.54***	−0.46***	0.09
Plot Edge Index	0.22*	−0.02	0.11	−0.47***	−0.28*	−0.22*	−0.30**
Presence							
Plot Size	0.31**	0.01	0.32**	0.06	−0.01	0.05	−0.02
Plot Shape	−0.07	0.14	0.09	0.05	0.10	0.17	0.23*
Plot Edge Density	0.26*	0.25*	0.33**	−0.23*	−0.50***	−0.48***	0.26*
Plot Edge Index	0.46***	0.30**	0.51***	−0.10	−0.29	−0.18	0.31**

Edge density and the edge index combine all edge types, whether occurring along the perimeter or internally, identified on 1:2000 aerial photograph-derived base maps of each plot. Data are from Keller (1990) and Keller et al. (2003)
[a] Species abbreviations as in Table 6.1
[b] Size (ha) of each of the 23 plots
Plot Shape = Perimeter length/2 × (Pi × Area) $^{1/2}$; equivalent to FRAGSTATS patch shape index (after Patton 1975)
Plot Edge Density = Total edge length (boundary and interior) on the plot/Area; equivalent to FRAGSTATS landscape edge density
Plot Edge Index = Total edge length (boundary and interior) on the plot/(Area$^{1/2}$); approximately equivalent to FRAGSTATS landscape shape index (LSI)
* $p < 0.05$, ** $p < 0.01$, *** $p < 0.001$

Table 6.4 Simple linear correlations (r) between bird density and (1) species-scaled, patch-specific landscape metrics, (2) ground-based explanatory variables, and (3) combinations of the two (see Tables 6.1 and 6.2)

Variable[b]	Species[a]						
	ALFL	HOWR	NAWA	CSWA	BTBW	BAWW	INBU
DEAC	0.42				0.68	0.53	
DEAC ÷ PS						0.27**	0.63 0.67[c]
DEACLA	0.43		0.44		0.91	0.72	
DEACLA ÷ PS		0.47 0.46[d]	0.44 0.42[d]		0.75	0.59	0.62 0.61[c]
DEAC_LALOWMID					0.75		
DEACLAratio	0.46		0.41				
DEACLAratio ÷ PS	0.21*						0.64 0.68[c]
DIST	−0.26**						
EIDCLA	0.50		0.38[f]		0.89	0.65	
EIDCLA ÷ PS			0.40 0.42[d]				
EIDEAC	0.50	.	0.38[f]		0.79		
EIDCLAratio	0.48		0.37[f]				
LA1M		0.42				0.46	0.38
LA1Mratio	0.38						0.43
LALOW				0.74		0.46	
LALOWratio		0.42	0.40	0.50			0.39
LALOWMID				0.59	0.59	0.61	
LALOWMIDratio			0.33		−0.44	−0.45	
LAMID		−0.35					−0.46
LAMIDratio	0.04 NS					0.35	−0.37
LAHIGH	−0.20*	−0.30**	−0.26**	−0.37			−0.47
MDC ÷ PS		0.44[e]	0.45	0.71			0.74[e]
MDCLA ÷ PS		0.53[e]	0.46[e]	0.83			0.66[e]
MDC_LAMID			0.32[e]				
MDCLAratio ÷ PS		0.49[e]	0.42[e]				0.71[e]
NUMHAB	0.36				0.80	0.68	
PLOTEI	0.50		0.4[f]		0.91	0.83	
VMRI		0.47[g]	0.46[f]		0.44		

◀ Single correlations with patch-specific variables (e.g., DEAC, NUMHAB) for HOWR, NAWA, and INBU are with Patch Type 9 (Shrub-opening edge), except as noted. Unless otherwise noted, all other single correlations with patch-specific variables are with the Patch Type (T) for which the species is listed as a Principal Associate in Table 6.1. Other than some variables included in MAXR regression models and shown here for illustrative purposes, only highly significant ($p < 0.001$) correlations are included

[a] Species abbreviations as in Table 6.1

[b] Variable definitions are found in Table 6.2

[c] Correlations for patch types 9 and 10, respectively. See Table 6.1

[d] Correlations for patch types 9 and 14, respectively. See Table 6.1

[e] Correlation for Patch Type 4. See Table 6.1

[f] Correlation for Patch Type 10. See Table 6.1

[g] Correlation for Patch Type 12. See Table 6.1

*$p < 0.05$, ** $p < 0.01$

NS Not Significant

Similarly, the negative correlations in Table 6.3 of Plot Edge Density and/or Plot Edge Index with three (BAWW, HOWR, INBU) of the remaining five edge species are misleading, and as with the BTBW, suggest they are not edge-associated. Yet, all three are more strongly positively correlated with species-specific edge types in Tables 6.4 and 6.5, and any student of bird study in the northeast would associate the occurrence of breeding individuals of these species with interspersed landscape component types (i.e., edge habitats) in the field. Similarly misleading negative correlations of edge species with more general and/or landscape scale (i.e., ecotone) edge measures can be found elsewhere (cf. INBU in Howell et al. 2008, Table 3; Connecticut Warbler (COWA) in Lapin et al. 2013, Table 1).

Also note that as observed by Cushman and McGarigal (2004), the choice of response variable, in this case density versus presence, influences the observed bird–habitat relationships. Here, both the composition of the most highly correlated explanatory variables and the strength of their correlations differ not only between species, but also between the two response variables for the same species. For example, as discussed previously, MDC and diameter of the equivalent area circle (DEAC) are good measures for determining the threshold size at which a species first appears and thus are good correlates of species' presence (Table 6.5). However, they are much less strongly correlated with density (Table 6.4 and Keller 1986). This is because both variables are patch type specific, and thus, generally independent of plot size (i.e., the amount of species-specific patch [MDC or DEAC] may occupy all or only a subsection of the plot [biotope(s)]). Density, however, is a direct function of plot size. Therefore, when exploring correlations of bird *density* with MDC or DEAC, it is reasonable to correct these two explanatory variables for plot size. Table 6.4 includes MDC, DEAC, and related variables that are corrected for plot size (e.g., MDC1 ÷ Plot Size). One or more of these variables were highly correlated ($p < 0.001$) with the density of all but the ALFL (Table 6.4) but were rarely as highly correlated with a species' presence (Table 6.5) as were uncorrected MDC- or DEAC-related variables. Differences in the explanatory strength of other variables are also evident and reinforce Cushman and McGarigal's (2004) point about the influence of response variable selection on development of species–habitat models.

Table 6.5 Simple linear correlations (r) between bird species presence and (1) species-scaled, patch-specific landscape metrics, (2) ground-based explanatory variables, and (3) combinations of the two (see Tables 6.1 and 6.2)

Variable[b]	Species[a]						
	ALFL	HOWR	NAWA	CSWA	BTBW	BAWW	INBU
DEAC	0.52		0.60 0.54[c]		0.79	0.57	0.63 0.66[c]
DEACLA	0.54	0.54 0.51[c]	0.58 0.55[c]		0.91	0.40[d] 0.74	0.56 0.58[c]
DEACLAratio	0.55	0.51 0.47[e]	0.64				0.51
DIST	−0.30**			−0.45			
EIDCLA	0.61	0.53	0.63 0.62[c]		0.90	0.76	0.53
EIDCLAratio		0.47	0.62				0.48
EIDEAC	0.57				0.85		
LA1M	0.39	0.46		0.61		0.04 NS	0.53
LA1Mratio	0.46	0.45	0.52	0.62			0.61
LALOW				0.76		0.49	0.34
LALOWratio		0.47	0.46	0.68			0.34
LALOWMID				0.44	0.49	0.59	
LALOWMIDratio			0.36			−0.33	
LAMID		−0.44			0.35		−0.72
LAMIDratio	0.05 NS	−0.34				0.35	−0.54
LAHIGH	−0.26**	−0.39	−0.35	−0.53			−0.77
MDC		0.54[f]	0.61[f]	0.70			0.68[f]
MDCLA	0.45[f]	0.57[f]	0.55[f]	0.65			0.56[f]
MDCLAratio		0.55[f]	0.58[f]				0.54[f]
NUMHAB	0.51	0.47	0.55 0.59[c]		0.85	0.75	0.63
PLOTEI	0.49	0.53[g]	0.56 0.56[h]		0.91	0.84	0.53 0.55[c]
VMRI		0.39[g]	0.48[g]		0.56		0.40

Single correlations with patch-specific variables (e.g., DEAC, NUMHAB) for HOWR, NAWA, and INBU are with Patch Type 9 (Shrub-opening edge), except as noted. Unless otherwise noted, all other single correlations with patch-specific variables are with the Patch Type (T) for which the species is listed as a Principal Associate in Table 6.1. Other than some variables included in MAXR regression models and shown here for illustrative purposes, only highly significant ($p < 0.001$) correlations are included

[a] Species abbreviations as in Table 6.1

[b] Variable definitions are found in Table 6.2

[c] Correlations for patch types 9 and 14, respectively. See Table 6.1

[d] Correlation for Patch Type 12. See Table 6.1

[e] Correlations for patch types 9 and 10, respectively. See Table 6.1

[f] Correlation for Patch Type 4. See Table 6.1

[g] Correlation for Patch Type 10. See Table 6.1

[h] Correlations for patch types 10 and 14, respectively. See Table 6.1

** $p < 0.01$

NS Not Significant

Lastly, note the strong negative correlations between high canopy (i.e., >7 m) leaf area (LAHIGH) and the five species (ALFL, CSWA, HOWR, INBU, NAWA) generally recognized as being associated with early successional forests and/or mid-stage oldfields undergoing woody invasion. As we noted earlier (Sect. 5.3 and Appendix B), such negative correlations of a species with habitat structure often only characterize where species generally do not occur, not specifically where they do occur, and thus may provide little information on how to manage the species. In this case, avian ecologists know that these species do not occur in closed canopy forests where values of LAHIGH are greatest. The more specific question is, "in what type of *open* canopy settings do they occur most frequently?" We, therefore, did not include LAHIGH in the list of variables entered into multivariate analyses for each of these species but instead generally focused on variables that described *early successional* vertical structure and component (e.g., sapling, shrub, grass) spatial arrangement (Tables 6.1 and 6.2, Appendix E).

Regression Analysis

Multivariate habitat models for each species were produced using MAXR multiple regression analysis as in Keller (1990) for both the generalized landscape metrics and the more species-specific explanatory variables (see Appendix E for a list of the latter variables tested with each species).

Per Keller (1990), the criterion was established that all explanatory variables retained in the selected model had to explain a significant ($p < 0.05$) portion of the variance in the data. Mallow's $C(p)$ statistic of fit, which is included with the SAS output (SAS 2011), was used to corroborate selection of the most parsimonious model for each species. $C(p)$ has recently been shown to be equivalent to the Akaike information criterion (AIC) for linear regression (Boisbunon et al. 2013) and thus incorporated a desirable information theoretic approach to model selection.

Models using the exclusively GIS-based general landscape metrics though, in general, highly significant (10 of 14 models $P < 0.007$, Tables 6.6 and 6.7) exhibited modest explanatory power ($R^2 = 0.022$–0.351) averaging $R^2 = 0.194$ for bird density and $R^2 = 0.182$ for species presence. Though not explaining consistently large portions of the variance in bird density or presence, nonetheless, several models (e.g., INBU, BTBW, BAWW density) incorporating these general landscape measures exhibited higher regression coefficients than did many studies we reviewed earlier. We attribute this primarily to the higher resolution imagery employed here, which allowed quantification of within-biotope (within-plot) edges not measurable at lower resolution. However, as noted in discussing the univariate correlations, many of the species' associations with the general edge metrics were misleadingly negative, which we again suggest is likely due to each of these metrics inclusion of the total length of all edge types that occurred on the plot, since both resolution and ecological scale are constant in both analyses.

In contrast, regression coefficients for MAXR models of density at the same resolution but using the more species-specific explanatory variables (Table 6.8) ranged from $R^2 = 0.377$ to 0.967 and averaged 0.646. Model regression coefficients for presence of the seven species (Table 6.9) ranged from 0.521 to 0.952 and averaged

Table 6.6 General landscape metrics included in models for the density of seven species of breeding birds at the Connecticut Hill WMA using MAXR Regression

ALFL[a] (0.114, $p = 0.023$[b])	HOWR (0.022, $p = 0.195$)	NAWA (0.056, $p = 0.0362$)	CSWA (0.219, $p < 0.0001$)
-Plot Shape[c]	Plot Edge Density	Plot Edge Density	-Plot Edge Index
Plot Edge Index			

BTBW (0.338, $p < 0.0001$)	BAWW (0.256, $p < 0.0001$)	INBU (0.351, $p < 0.0001$)
-Plot Edge Density	-Plot Edge Density	Plot Edge Density
-Plot Edge Index	-Plot Edge Index	-Plot Edge Index
-Plot Size	-Plot Size	-Plot Size
		Plot Shape

Variables are listed in order of decreasing variance (F value) explained within the model. R^2 values are shown in parentheses
[a] Species abbreviations as in Table 6.1
[b] $p =$ significance of the model. All included variables were $p < 0.05$
[c] Variable definitions are found in Table 6.2

Table 6.7 General landscape metrics included in models for the presence of seven species of breeding birds at the Connecticut Hill WMA using MAXR Regression

ALFL[a] (0.228, $p < 0.0001$[b])	HOWR (0.091, $p = 0.007$)	NAWA (0.279, $p < 0.0001$)	CSWA (0.054, $p = 0.039$)
Plot Edge Index[c]	Plot Edge Index	Plot Size	-Plot Edge Density
-Plot Shape		Plot Edge Density	

BTBW (0.251, $p < 0.0001$)	BAWW (0.235, $p < 0.0001$)	INBU (0.133, $p = 0.0044$)
-Plot Edge Density	-Plot Edge Density	Plot Shape
		Plot Edge Density

Variables are listed in order of decreasing variance (F value) explained within the model. R^2 values are shown in parentheses
[a] Species abbreviations as in Table 6.1
[b] $p =$ significance of the model. All included variables were $p < 0.05$
[c] Variable definitions are found in Table 6.2

0.687. Compare these, for example, with models for shared or ecologically similar species in Lichstein et al. (2002b), Betts et al. (2006), and Howell et al. (2008).

The principle differences between these models and those employing more general measures of GIS-based landscape attributes or on-the-ground vegetation structure are the inclusion here of variables that for each plot quantify (1) the largest size circle (MDC/DEAC) containing the greatest density of that combination of landscape components thought to be associated with the species of interest and (2) the actual amount of LA within specified vertical intervals observed to be used by each species within the landscape.

Table 6.8 Species-scaled variables included in models for the density of seven species of breeding birds at the Connecticut Hill WMA using MAXR Regression

ALFL[a] (0.337)	HOWR (0.414)	NAWA (0.421)	CSWA (0.716)
EIDCLA10[b,c]	MDCLA4 ÷ PS	MDCLA4 ÷ PS	MDCLA1 ÷ PS
-DIST10	VMRI12	DEACLAratio9	LALOW
DEACLAratio10 ÷ PS	-LAMID	MDC4LAMID	
LAMIDratio		LALOWratio	
		EIDCLA14 ÷ PS	

BTBW (0.967)	BAWW (0.914)	INBU (0.715)
DEACLA13	DEACLA13 ÷ PS	-LAMID
EIDCLA13	DEACLA13	DEACLAratio10 ÷ PS
DEAC13LALowMid	NUMHAB13	DEACLAratio9 ÷ PS
DEAC13	PLOTEI13	MDC4 ÷ PS
EIDEAC13	DEAC13	VMRI14
NUMHAB13	DEAC13 ÷ PS	
PLOTEI13	LALOW	
	LA1M	

Variables are listed in order of decreasing variance (F value) explained within the model. All models were highly significant ($p < 0.0001$). Model R^2 values are shown in parentheses and averaged 0.646
[a] Species abbreviations as in Table 6.1
[b] Variable definitions are found in Table 6.2
[c] All included variables were significant at $p < 0.05$

Variables containing MDC or DEAC were included in all models for each of the seven species (total of 14 models), and multiple variables containing MDC and DEAC were included in 9 of the 14 models. All models contained multiple measures that included LA. The overall strength of the models (R^2 14 models = 0.667) suggests the potential utility of such variables in predicting both threshold occurrence (presence) and density for species of interest. Additionally, the covariates are compositionally and/or structurally specific enough (i.e., they reference individual edge types and height intervals) to provide practical guidance for development of management prescriptions and conservation strategies for species of interest.

From an interpretability standpoint, we recognize that at first glance, some combination variables included in the models may appear unduly complex. However, decomposition of each variable into its component parts reveals the underlying biological relationship of the variable to the species with which it is associated (Figs. 6.1 and 6.2). For example, one of the highest univariate correlates of presence of the NAWA, EIDCLAratio9 ($r = 0.62$), consists, for each plot, of the diameter of the circle with the highest density of shrub-opening edge (DC = DEAC of Patch Type 9) multiplied by (1) the actual density of Patch Type 9 edge within that circle (EI) and (2) the ratio (value 0–1) of the observed LA in the 0-3 m vertical interval to the median LA in that interval for all plots in the

Table 6.9 Species-scaled variables included in models for the presence of seven species of breeding birds at the Connecticut Hill WMA using MAXR Regression

ALFL[a] (0.552)	HOWR (0.521)	NAWA (0.546)	CSWA (0.673)
EIDCLA10[b,c]	PLOTEI10	NUMHAB14	LALOW
-DIST10	LALOWratio	VMRI10	MDCLA1
LAMIDratio	-LAMIDratio	NUMHAB9	LALOWMID
PLOTEI10	DEACLAratio9	LA1Mratio	MDC1
NUMHAB10	EIDCLA9	EIDCLAratio9	LA1M
LA1Mratio	DEACLAratio10		
	DEACLA14		

BTBW (0.947)	BAWW (0.800)	INBU (0.764)
DEACLA13	NUMHAB13	-LAMID
EIDEAC13	DEACLA12	DEAC14
EIDCLA13	DEAC13	PLOTEI9
DEAC13	LA1M	LALOW
VMRI13		LA1M
NUMHAB13		LA1Mratio

Variables are listed in order of decreasing variance (F value) explained within the model. All models were highly significant ($p < 0.0001$). Model R^2 values are shown in parentheses and averaged 0.687

[a] Species abbreviations as in Table 6.1
[b] Variable definitions are found in Table 6.2
[c] All included variables were significant at $p < 0.05$

study area, a calculation that weights most heavily an intermediate amount of LA in the interval (see exact calculation in Table 6.2). Thus, the variable combines measures of horizontal and vertical vegetation structure (i.e., the foliage profile *sensu* MacArthur et al. 1962) with a measure of the profile's spatial extent (optimally shaped size) and quality (i.e., the area of leaves within it) on a given plot. This is simply a more quantitative, three-dimensional extension of MacArthur's patch (foliage profile) concept of species occurrence, the original quantification of which was based solely on proportions of foliage in three vertical intervals (op. cit.).

Similarly, other combination variables produce metrics that quantify various aspects of structure (e.g., LALOWMID) and the degree of clustering (e.g., VMRI, DEACLA) of landscape components hypothesized as relevant to a particular species' or assemblage's use of the landscape. In comparison, consider, for example, the edge associations for BTBW at the territory scale evaluated here (e.g., DEACLA13) versus associations of this species with more general measures of structure that do not consider spatial arrangement (e.g., sapling density and basal area of conifers) at both similar scales and a more extensive scale in Smith et al. (2008). The use of less spatially explicit non-edge variables results in a different interpretation of the primary scale of habitat selection for this species (op. cit.).

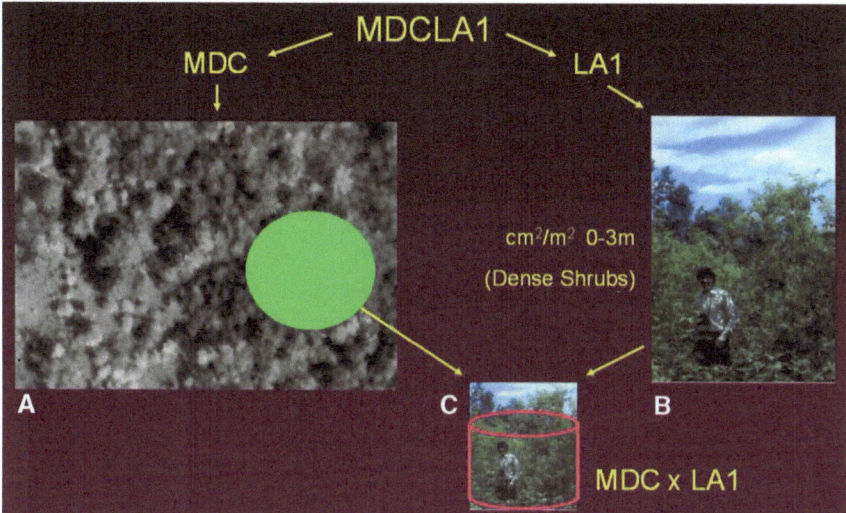

Fig. 6.1 The variable MDCLA1 combines *A* the maximum diameter circle in a dense shrub patch (Patch Type 1) with *B* the leaf area in the 0–3 m interval above the ground within that patch to produce *C* a three-dimensional (cylindrical) descriptor of the functional size of the dense shrub patch with energetically optimal shape. This variable was highly correlated ($r = 0.65$, $p < 0.001$) with the occurrence of the CSWA, a low-foliage-gleaning insectivore of early successional forests and oldfields. When corrected for plot size, it was even more highly correlated with CSWA density ($r = 0.83$, $p < 0.001$)

Although many additional comparisons can be made between the general metric and species-specific variable models, we shall touch on only a few. First, only 4 of 14 models employing the general metrics included plot size. This might lead to the erroneous conclusion that habitat or patch size is unimportant, even at the small size of many of the plots studied (16 of 23 plots <10 ha). Yet, as noted, all 14 models based on more species-specific explanatory variables included measures of patch size, and 11 of those models included multiple measures of size. This strongly suggests that at or near threshold occupancy levels, it is not simply the size of the plant community or biotope in which the species is found that is important to its occurrence but rather the *functional* size of the habitat available *within* that biotope, a measurement that in GIS-based analyses can only be obtained when image resolution is appropriate to the focal taxon (Huston 2002). Here, we measured functional habitat size as either the largest optimally shaped patch (DEAC, MDC) within the landscape examined or by otherwise quantifying the degree of clustering (e.g., VMRI) of landscape components (i.e., patch types) specific to the species of interest.

Second, Huston (2002) discussed the potential confusion in species–habitat modeling of the unimodal ("humpbacked") response curve, where a response variable first increases at low levels of an environmental variable and then decreases at higher levels (cf. Van Horne 2002, Fig. 4.2) leading to what he described as response reversals. This particular curvilinear response describes the relationship

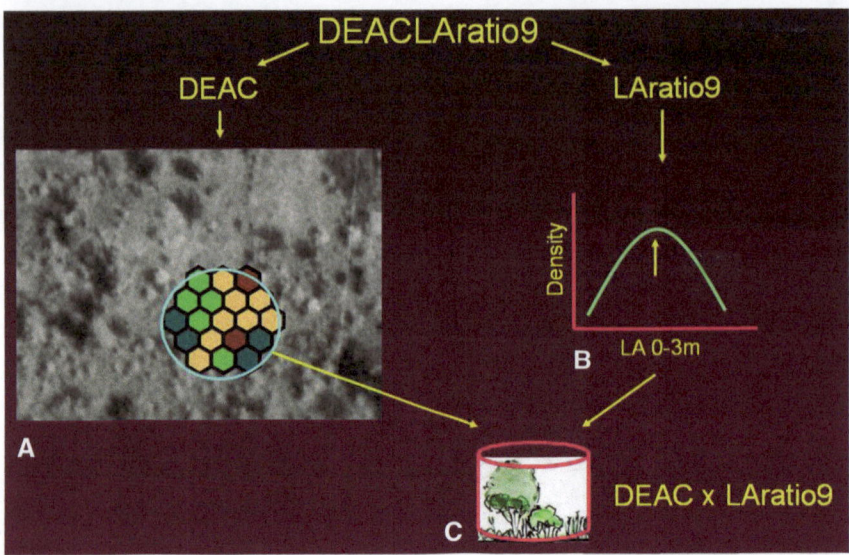

Fig. 6.2 The variable DEACLAratio9 combines *A* the diameter of the equivalent area (to MDC) *circle* having the highest density of shrub-opening edge (DEAC of Patch Type 9) with *B* the ratio (value 0–1) of the observed leaf area (LA) in the 0–3 m vertical interval to the median LA in that interval for all plots in the study area, a calculation that weights most heavily an intermediate amount of LA in the interval (see exact calculation in Table 6.2). This produces *C* a three-dimensional (cylindrical) descriptor of the functional size of the shrub-opening edge patch with energetically optimal shape. This variable was highly correlated with the presence of the HOWR, NAWA, and INBU ($r = 0.51 - 0.64, p < 0.001$)

between many edge species of birds and LA within the height interval they occupy (Fig. 6.2B and J. K. Keller, unpubl. data). Ecologically, the apparent preference for an intermediate amount of LA within the height interval used for foraging and/or nesting by an edge species is quite logical. Too few leaves may result in fewer prey items and/or locations to conceal a nest. Too many leaves, often an indication of a more advanced stage of succession, result in a lack of the openings favored or required (e.g., ALFL) by such species. Spatially, intermediate levels of LA within edge habitats typically reflect the juxtaposition of landscape components containing high densities of leaves at a species-specific height with components that lack leaves at the same height (i.e., an edge) such as between shrubs and open grass (Fig. 6.2C). To transform this curvilinear relationship with leaf area into a habitat descriptor suitable for inclusion in linear regression, we developed the variable LAratio (Table 6.2), which weights intermediate levels of LA most highly (see Rehm and Baldassarre 2007 for a similar variable applied to open water-marsh edge). In our analysis, all but one of the 8 models for the ALFL, HOWR, INBU, and NAWA included one or more variables with LAratio (Tables 6.8 and 6.9), suggesting the importance to these edge species of an intermediate amount of LA within their most-used height interval. Variables combining leaf area or other vertical spatial distribution of landscape elements with horizontal measures derived from GIS, such as quantified by LAratio, are largely absent from current studies but could presumably be generated using LiDAR combined with GIS metrics of spatial arrangement.

We also observed that even at the high resolution employed here, selection of a relatively large grid cell size (i.e., relative to GRD), in this case 100 m^2 on the ground, resulted in a fair amount of within-cell heterogeneity. Such cell-size-based heterogeneity was also a feature common to the lower resolution studies we reviewed; although in the reviewed studies, the absolute sizes of the mapping units were several orders of magnitude larger than our 100-m^2 grid cells. A quick examination of the hexagonal-celled overlay in Fig. 2.1, for example, shows the inclusion of deciduous shrubs (or sapling conifers) and grass within some individual cells. These cells, known in remote sensing as "mixed pixels," were neither predominantly grass nor shrubs and were classified as landscape component type 2 (deciduous shrub–grass) or 3 (conifer sapling–grass, Appendix C). Unlike true mixed pixels in remote sensing, however, where multiple *unresolved* elements (L-resolution) with different reflectance characteristics in a single pixel often result in misclassification of that pixel, the use of stereoscopic, H-resolution photos in this analysis allowed accurate classification of cells with multiple elements to structurally consistent component types. Component types 3, 4, and similar cell types were grouped and defined as Patch Type 4, open shrubs (Appendix D), because of the association of several shrubland species with clusters of these component types.

Thus, the shrub-opening habitat type, although a *mix* of grass and shrubs containing extensive edge, is quantified in the GIS at the 100 m^2 cell size as if it were a "solid" patch type (i.e., composed of clusters of a single component such as all grass). As a result of classifying these heterogeneous cells as a single combined type (=open shrubs), the size of this patch type is measured as a MDC (Fig. 3.2) within a cluster of shrub–grass cells rather than as a DEAC (Fig. 3.5) of the edges between adjacent cells classified separately as "shrub" or "grass." This explains the inclusion of variables containing MDC4 in models for edge species such as the HOWR and NAWA. These species are indeed associated with edges; however, in this case, the edges, although identifiable in the image, are more fine-grained than the mapping units (i.e., occur *within* the cells), resulting in their association with a non-edge metric.

Though all of the species-specific variable models are highly significant, some clearly have greater explanatory power than others. In particular, models of density for the ALFL, HOWR, and NAWA, although reasonably good (average $R^2 = 0.404$), lagged those for the other four species and exhibited substantially lower regression coefficients for density than for presence of the same three species (average R^2 presence = 0.540). Among many possible reasons for the lower explanatory power of these three density models, species natural history may play a role. The HOWR, for example, nests on or near the ground in brush piles as well as in cavities, both of which represent somewhat specialized nest sites. Thus, nest site availability may limit its presence even when otherwise appropriate habitat appears to be available.

The ALFL, in contrast, was observed to be temporally inconsistent in its use of particular plots, being present in alternate years or for only a single year in otherwise apparently suitable settings (Keller 1990). Models for both ALFL density and presence included and were negatively correlated with the distance to the nearest location where other males of the species were observed holding a territory (the variable DIST10). This suggests there may be a degree of social facilitation in this species use of a particular site that is independent of habitat availability (see Fig. 2 in Keller 1990; Campomizzi et al. 2008). Social facilitation is one of the processes cited as causative of spatial autocorrelation and observed reductions in explanatory power for models that are based primarily on habitat structure, such as those included here (see discussion in Sect. 5.4). In this case, removal of DIST10 from the regression for ALFL density results in a MAXR model that explains only 28 % of the variance in the data set versus 38 % when plot insularity is considered, a decrease of more than one-fourth of the model's explanatory power.

In addition to the use of species-specific variables measured on a high-resolution image, another aspect of this analysis worth noting is the use of spot-mapping. We believe this survey method was instrumental in helping to identify useful explanatory variables because it allowed better researcher assessment/ understanding of (1) the relationship of territory locations (i.e., areas used) to the structure of vegetation within them, (2) that structure's spatial distribution on the landscape, and (3) other attributes of the space occupied by the species of interest such as slope aspect, vegetation composition, and site context than could be obtained from typically less frequent or less intensive point count surveys. Similarly, radiotelemetry studies (e.g., Leonard et al. 2008) can provide more precise information on space use by individual species. Although more time consuming, we suggest these types of more intensive investigations should be given serious consideration in future studies of avian species–habitat relationships.

Lastly, in an alternate test of the predictive power of such species-specific variables, Keller (1990) also developed stepwise discriminant function analysis (SDFA) models of presence/absence for the ALFL, BAWW, and BTBW included here. Using a data splitting/model building procedure similar to that employed by Howell et al. (2008) and essentially the same explanatory variables illustrated above, the average classification accuracy for jackknifed SDFA models and tests of models was 92.1 and 87.1 %, respectively, for presence of the three species. Compare this with the approximate 60 % model accuracies achieved by Howell et al. (2008) and Thogmartin et al. (2004b), for example, who (1) employed Landsat data, (2) did not include any explanatory variables that measured spatial arrangement or edge, and (3) used minimum analytical units of 5.76 ha, and 800 ha, respectively.

Despite the explanatory power of the species–habitat models developed here, as noted by Keller (1990), they are likely not the best models for these species, even within the context of this data set. For example, analyses of additional curvilinear rather than simple linear relationships, the inclusion of additional variables relating to site context and composition of the surrounding matrix, and/or inclusion of a measure of species detectability (MacKenzie et al. 2002) might well further improve their explanatory power.

6.1 Summary

At the outset of our discussion, we identified a number of issues that we and other authors (particularly Huston and others in Scott et al. 2002) viewed as problematic for GIS-based analyses of species–habitat relationships. Huston (2002: 11) summarized these issues stating *"Unless ecological patterns, and the environmental conditions that influence the processes that determine the patterns, are quantified appropriately, species occurrences, patterns of species distributions, and variation in population viability cannot be understood or predicted. The issues of sample resolution, sample density, and size (or duration) of a study must be adequately addressed before it is possible to quantify and understand the complex interactions of ecological processes."*

In exploring these limitations, both on the use of GIS and on research methods in general, we examined how initial researcher choices of image resolution, scale(s) of analysis, explanatory variables—including landscape metrics—and location and area of samples can influence analysis results, interpretation, predictions, and study-derived management prescriptions and/or conservation strategies. Although the generally low predictive power of current GIS-based species–habitat models results from combinations of these factors, the choice of insufficiently high-resolution imagery appears to be the most consistently limiting element in these studies.

Despite universal agreement among numerous authors about the need for matching scale of analysis to the scale of organism habitat use (e.g., Wiens 1989b; Huston 2002; O'Conner 2002; Trani 2002; Van Horne 2002), we found limited evidence that this seemingly inarguable tenet is applied rigorously in GIS-based studies. Instead, researchers appear to rely on readily available imagery without considering critically the consequences of image resolution for their particular study (McDermid et al. 2009). Regardless of the selected imagery, the choice is either unexplained or argued as "reasonable" for the taxon under study, even though, particularly in the case of passerines and other smaller animals, the image resolution (i.e., the grain of analysis) is often low and either the minimum mapping unit (mmu) or minimum sampling unit large relative to space use by the focal taxon (e.g., Betts et al. 2003; Lawler et al. 2004; Thogmartin et al. 2004b; Howell et al. 2008; Cornell and Donovan 2010; LeBrun et al. 2012). In the former case, although by definition classified as a single type, mmu's may actually be composed of multiple landscape component types, among which there may be subsets of types (i.e., the true grain) that are important to habitat selection by the focal taxon (Fig. 2.1). Yet, at inappropriately low resolution, these individual elements are indistinguishable and thus unmeasurable (Fig. 1.1). This precludes analysis of local or within-territory heterogeneity that may critically influence species occurrence. Alternately, or additionally, extensive minimum sampling areas (e.g. Thogmartin 2004b; Lebrun et al. 2012) may include only small areas of habitat within them that is then obscured statistically by the average conditions of the entire area (Huston 2002). Under either of these scenarios, resolution-limited analyses may produce (1) low or nonsignificant correlations with ecologically meaningful and potentially managerially useful predictor variables, (2) strong correlations with ecologically less meaningful variables, and/or (3) completely misleading (e.g., inverse) correlations. Various combinations of these latter points and the need for higher resolution imagery are repeatedly identified post hoc in the Discussion sections of papers we reviewed.

However, even with more taxon-specific, higher resolution imagery, variables such as proportions and indices (of biotope-level plant associations) typically found in recent studies and represented by the general landscape metrics tested here, simply do not quantify adequately either the patch type (i.e., solid or edge, and composition) or the critical, clustered nature of threshold-sized parcels of habitat that foster the appearance of the first breeding pair of a particular species within a plant community or other biotope.

In our demonstration analysis of the habitat relationships of seven species of passerines, we attempted to address the issues raised throughout our discussion (Chaps. 2–5). Specifically, we

1. Employed high-resolution stereoscopic aerial photography to identify and classify *landscape component types* (Appendix C) at an ecological scale relevant to space use by the passerines under study;
2. Used spot-mapping to closely match detection locations and scales of bird–habitat use to vegetation and GIS samples;
3. Identified exploratory, species-specific *patch types* (i.e., combinations of solid or edge landscape components) that represented hypothesized habitat composition and architecture for the species under consideration (Appendix D);
4. Applied several novel, biologically intuitive covariates that quantify the functionally largest (based on energetically optimal territory shape) example of a specified solid or edge patch type within an area of study; and
5. Developed a new variable (LAratio) that addresses a curvilinear (unimodal) response of species to increase in a particular resource. In this application, LAratio applied to edge species of early successional insectivorous birds that favor intermediate amounts of leaf area within selected vertical intervals.

Overall, MDC, DEAC, and VMRI, in combination with the vertical distribution of LA, distinguish threshold and larger size aggregations of species-specific habitat where corresponding species populations occur within a landscape. The application of MDC and DEAC to identification of thresholds is consistent with O'Conner's (2002) proposed *constraints* paradigm, which accounts for a species' local carrying capacity by identifying "the most severe of multiple alternative limits." In this case, this *most severe limit* is the threshold *size* of species-specific MDC or DEAC at which the first breeding pair of a species appears on the landscape (Keller 1986, 1990). However, unless habitats and the biotopes within which they are embedded are viewed and analyzed at an image resolution and under an accompanying classification system that closely match the scale of habitat selection by the focal taxon [i.e., the domain of resolution (Sect. 2.5.1 and Fig. 2.1)], these thresholds (limits) may be missed because the landscape appears increasingly homogeneous at coarser image resolutions (Fig. 1.1).

Finally, in his essay on critical issues for improving ecological predictions, Huston (2002:11) concluded that "*Hypotheses that are sufficiently robust to avoid definitive falsification not only provide a statement of our understanding of particular ecological processes, but also improve predictions by identifying the critical independent variables and describing their effect on ecological properties.*" We submit that MDC, DEAC, and LAratio represent such variables and that their application in the context of the modeling approach we have described will improve future predictions of species distributions and management of their populations. Although the models we developed certainly can be improved, the approach to GIS-based species–habitat analyses we demonstrate serves to highlight the three main recommendations of our discussion: (1) select the resolution of remotely sensed imagery to match the scale at which the focal taxon uses the landscape, frequently a body-size-related attribute, (2) identify measures (explanatory variables) of habitat architecture (vertical and horizontal structure), size, configuration, quality, and context that reflect the way the focal taxon uses the subset of the landscape it occupies, and (3) as closely as possible, match habitat samples to areas actually *occupied* by the focal taxon when a sample is classified as being where the organism is "detected" or "present."

6.2 Recommendations

Based on the results of the analysis in Chap. 6, we suggest the approach we have outlined has application to management and conservation of a variety of species. Models with good predictive power that are based on biologically interpretable variables related to species use of the environment suggest practical management prescriptions and conservation strategies. These prescriptions and strategies are in turn more defensible because of the strength of the analyses and the underlying ecological intuitiveness of the models. Our general recommendations are summarized as follows:

• If the research goal is species or taxon management, conduct analyses at the species [e.g., Golden-winged Warbler (GWWA)] or functionally similar assemblage (e.g., low-gleaning solid patch insectivores) organizational level(s) rather than at the level of broader assemblages (e.g., all early successional neotropical warblers) or communities (Appendix B). Data can be combined and broader inferences (e.g., community richness) made later, if this is of interest (see also Morrison and Hall 2002: 49 for recommended study standards).

- Using a standard scientific approach (i.e., a review of natural history data and prior studies combined with field observation of how the focal taxon uses the landscape), ask the following:

  - What is the size of a typical breeding territory or home range? If examining assemblage subsets (e.g., the foliage gleaners cited above), what is the range of territory sizes under study?
  - Within the territory or home range, what subset of the available vegetation or landscape structure is used (e.g., for foraging, nesting, denning, breeding display) and over what seasonal or successional time frame, if any? Consider the three-dimensional (i.e., both vertical and horizontal) use of the space.
  - What heterogeneity exists within this subset (i.e., the habitat) that may be of interest, or essential, to quantify? How many potentially meaningful landscape components and/or edge types are found within the territory?
  - Are there aggregations or gradients of particular (solid) components or edge types consistently associated with the territory?
  - Will the remotely sensed imagery contemplated for use resolve landscape components and potential intraterritory heterogeneity that reflect the scale at which the species of interest selects and uses habitat?
  - Does the classification system proposed reflect the landscape component composition and heterogeneity described above?
  - If the scale of habitat use or association is unclear, are the smallest resolvable landscape components of potential interest in the imagery more than 5x − 10x the body length of the species of interest? Are there morphological or physiological attributes of the species that suggest the need for resolution of even smaller landscape components?

- Use the *function* of the focal taxon, a best estimate of the *scale* at which it uses the landscape, and the ecological time frame in which it uses the landscape to guide identification of more explicit landscape components and the appropriate image classification system.
- Use the same attributes (function, scale, successional time frame) to identify a series of local plots selected to sample a range of habitat conditions over which the focal taxon occurs (i.e., an environmental gradient) (Wiens et al. 1987; see also MacKenzie and Royle 2005 regarding design of companion occupancy studies).
- Measure *component and/or edge types, and their attributes, including energetically efficient (i.e., circle-based) spatial arrangements that could reasonably be interpreted as meaningful* (biologically relevant) to the focal taxon.
- Consider the composition of the surrounding landscape and quantify its extent and potential ecotone effects if it differs significantly from the study area. Explicitly consider the landscape *context* as well as its *content*.
- For GIS-based species–habitat analysis, always use the highest resolution imagery available and affordable. One can always aggregate information. One can never go back later and measure smaller components or their heterogeneity

on resolution-limited imagery (see McDermid et al. 2010 for an excellent discussion of an Application Framework for optimizing remote sensing data and methods in an ecological context).

• Keep both the variables, particularly GIS metrics, and statistical analysis as simple and biologically intuitive as possible (Stauffer 2002). As Murtaugh (2007:56) noted, *"Simple models are more likely to convey the key features of the data to other scientists, and they increase the chances that different analysts will reach the same conclusions."* De Knegt et al. (2010: 2463) concurred stating *"...even the most advanced and computer-intensive statistical procedures are no guarantee for improving our understanding of ecological responses, as such methods often do not give straightforward information about the underlying processes."*

• Conduct preliminary analyses of a representative subset of the data using both fine-grained and coarse-grained variables to determine which types of variables or combinations have better explanatory power (McDermid et al. 2010) and to assess the variance associated with each variable. If appropriate, use statistical inference to determine what sample sizes would be required to discern differences among samples at various confidence intervals (e.g., see Kuehl 2002).

Appendix A

Level	GRD (m)	Visible national imagery interpretability rating scale (NIIRS)—Civil
0	–	Interpretability of the image is precluded by obscuration, degradation, or very poor resolution
1	>9.0	Distinguish between major land use classes (e.g., urban, agricultural, forest, water, barren)
		Detect a medium-sized port facility
		Distinguish between runways and taxiways at a large airfield
		Identify large-area drainage patterns by type (e.g., dendritic, trellis, radial)
2	4.5–9.0	Identify large (i.e., greater than 160 acre) center-pivot irrigated fields during the growing season
		Detect large buildings (e.g., hospitals, factories)
		Identify road patterns, such as clover leafs, on major highway systems
		Detect ice-breaker tracks
		Detect the wake from a large (e.g., greater than 300 Ft) ship
3	2.5–4.5	Detect large-area (i.e., larger than 160 acres) contour plowing
		Detect individual houses in residential neighborhoods
		Detect trains or strings of standard rolling stock on railroad tracks (not individual cars)
		Identify inland waterways navigable by barges
		Distinguish between natural forest stands and orchards
4	1.2–2.5	Identify farm buildings as barns, silos, or residences
		Count unoccupied railroad tracks along right-of-way or in a rail yard
		Detect basketball court, tennis court, and volleyball court in urban areas
		Identify individual tracks, rail pairs, control towers, switching points in rail yards
		Detect jeep trails through grassland

(continued)

© The Author(s) 2014
J.K. Keller and C.R. Smith, *Improving GIS-based Wildlife-Habitat Analysis*,
SpringerBriefs in Ecology, DOI 10.1007/978-3-319-09608-7

(continued)

Level	GRD (m)	Visible national imagery interpretability rating scale (NIIRS)—Civil
5	0.75–1.2	Identify Christmas tree plantations
		Identify individual rail cars by type (e.g., gondola, flat, box) and locomotives (e.g., steam, diesel)
		Detect open bay doors of vehicle storage buildings
		Identify tents (larger than two persons) at established recreational camping areas
		Distinguish between stands of coniferous and deciduous trees in leaf-off condition
6	0.40-0.75	Detect narcotics intercropping based on texture
		Distinguish between row (e.g., corn, soybeans) crops and small-grain (e.g., wheat, oats) crops
		Identify automobiles as sedans or station wagons
		Identify individual telephone/electric poles in residential neighborhoods
		Detect foot trails through barren areas
7	0.20–0.40	Identify individual mature cotton plants in a known cotton field
		Identify individual railroad ties
		Detect individual steps on stairway
		Detect stumps and rocks in forest clearings and meadows
8	0.10–0.20	Count individual baby pigs
		Identify a United States Geological Survey (USGS) benchmark set in a paved surface
		Identify grill detailing and/or the license plate on a passenger/truck-type vehicle
		Identify individual pine seedlings
		Identify individual water lilies on a pond
		Identify windshield wipers on a vehicle
9	<0.10	Identify individual grain heads on small grain (e.g., wheat, oats, barley)
		Identify individual barbs on a barbed wire fence
		Detect individual spikes on railroad ties
		Identify individual bunches of pine needles
		Identify an ear tag on large game animals

Source Encyclopedia of optical engineering 2003

Appendix B

Response Variables: Individual Species Versus Assemblage Versus Guilds

Individual Species

There are many reasons for studying individual species–habitat relationships. Even within this elementary level, however, there is still a variety of response variables, such as presence (i.e., detected versus non-detected), density, relative abundance and reproductive success, available to investigators. Cushman and McGarigal (2004) examined how the choice of response variable, in this case relative abundance versus detection/non-detection, influenced the observed bird–habitat relationships. They found that patch (sensu plant community) and landscape level GIS variables were more important in explaining detection/non-detection data than abundance data, particularly for uncommon species. We suggest this is because GIS variables such as size and shape better characterize the spatial features of landscapes that are cues for patch colonization (i.e., thresholds) than do plot-level (local) variables, which in their study were more strongly associated with the abundance of more common species. Other studies have found similar disparities between alternative measures of species–habitat association (e.g., Lawler et al. 2004; Smith et al. 2007; see also Chap. 6 herein).

Game management and the development of conservation guidelines have long relied on studies of the habitat associations of focal species (e.g., Bump 1947; Nixon et al. 1988; Noss 1991; Zabel et al. 2003; Davis et al. 2007). One type of study in this realm attempts to develop management strategies for individual species by detecting thresholds in habitat amount using GIS metrics. Here, it is first essential to define accurately habitat association(s) for the species under consideration to ensure that landscape metrics are relevant to that species (Fischer et al. 2004). Although this is unambiguous in simulation models (e.g., Flather and Bevers 2002), it is quite challenging in empirical studies. In some cases, researchers simply have relied on qualitative definitions of what likely constitutes habitat (e.g., Homan et al. 2008) or have used general land cover classifications

© The Author(s) 2014

J.K. Keller and C.R. Smith, *Improving GIS-based Wildlife-Habitat Analysis*, SpringerBriefs in Ecology, DOI 10.1007/978-3-319-09608-7

[e.g., "forest" (Trzcinski et al. 1999); "native forest" (Lindenmayer et al. 2005)] despite the lack of detail in these labels.

Additionally, it is well recognized that, even within a given plant community such as a deciduous forest, species are typically associated most strongly with one structural aspect (subset) of the community or another: for example, red-eyed vireo (*Vireo olivaceous*) with canopy foliage, ovenbird with forest floor, Louisiana Waterthrush (*Parkesia motacilla*) with a stream in the forest, and flycatchers with canopy openings at various heights (cf. Holmes et al. 1979). Thus, certain variables used to describe the entire plant community within which a species resides may have little or no biological relevance to that species' distribution (Keller and Smith 1983; Rice et al. 1984) within the community.

Even when correlated with a species occurrence, only a portion of a variable's distribution may actually be influencing the correlation (Wiens 1989a). For example, percent forest cover is often associated with presence and abundance of the wood thrush (*Hylocichla mustelina*). However, studies have shown that, more specifically, it is the density of forest understory (i.e., in a subset of all forests) that most strongly influences wood thrush abundance within forests of appropriate size to support their populations (e.g., deCalesta 1994).

Lastly, thresholds in habitat amount appear to be more easily identifiable with GIS in agricultural (i.e., patchy, ecotone delineated) landscapes and island archipelagos than in forested landscapes (Mönkkönen and Reunanen 1999). However, this may be due, in part, to (1) the image resolution, (2) the spatial scale at which habitat is analyzed, and (3) the difficulty in correctly identifying the amount and distribution of habitat for individual species in more complex forest mosaics or other landscapes with spectrally indistinct landscape components (Smith et al. 2001). Boundaries between plant communities in the latter case are often indistinct (i.e., gradients), as opposed to well-delineated agricultural ecotones (Wiens 1994). We discuss the issues of image resolution, scales of analysis, habitat specificity, and quantitative description in Chaps. 2 and 3.

Assemblages

When exploring questions regarding habitat associations at higher organizational levels, such as causes of assemblage richness (e.g., an entire bird "community"), it is not necessarily most insightful to analyze the entire assemblage as a single unit [i.e., correlate measures of the entire assemblage (e.g., richness, diversity) with community- or landscape scale explanatory variables (Wiens 1989a; but cf. Matias et al. 2010)]. This is because of statistical and biological attributes of both the dependent and explanatory variables associated with broader organizational levels.

First, the range of independent variables often required to describe adequately the breadth of floristic and physiognomic structure of a landscape (i.e., multiple biotopes/communities) (1) can result in the inclusion of many variables that have little explanatory power for the distribution of particular species (see previous Chapter) and (2) may require collection of an inordinately large dependent data set to allow statistically valid examination of all possible variables used to describe the landscape. The use of PCA and other ordination techniques can reduce the number of variables to a statistically acceptable level. However, the multivariate axes generated by these techniques can also obfuscate species–habitat relationships to the point that interpretation of the results and development of specific management recommendations are difficult (Young and Hutto 2002; cf. Figs. 8, 9, and 10 in Cushman and McGarigal 2003; Figs. 1 and 2 in MacFaden and Capen 2002; and the description of L-PC2 on pg. 142 of Saab 1999).

Secondly, these difficulties may be especially magnified at the landscape scale where species composition includes an even wider range of functionally dissimilar species and structurally disparate plant communities. In this case, we would expect analyses of assemblage richness or assembly-wide habitat associations to be less likely to produce strong habitat correlations (cf. DeGraaf 1991) than analyses that consider both functionally similar species (i.e., guild members) and explanatory variables more specific to habitat selection mechanisms of such functionally related subsets of assemblages (cf. O'Connell et al. 1998, 2000). We suggest this is somewhat analogous to Diamond and Gilpin's (1982) well known "Owls vs. Hummingbirds" rebuttal to Conner and Simberloff (1979) regarding appropriate null models to test for competition among Pacific island bird species (i.e., all species are not functionally similar, should not be expected to compete with one another, and therefore should not be lumped together in the dependent data pool). These issues may be less problematic in linear landscapes (e.g., riparian corridors) because such landscapes frequently contain a restricted number of land use/cover types with a limited range of sizes and shapes, which reduces both the variety and range of values of the independent variables required to characterize the landscape adequately (e.g., Saab 1999).

Furthermore, although landscape or geographically more extensive analyses may produce strong correlations between individual species or suites of species and general habitat descriptors like percent cover, they may still provide insufficient information to manage species linked to these variables (e.g., Lawler et al. 2004). For example, in a landscape scale analysis of an entire assemblage, most early successional species would have strong negative correlations with forest cover, and many forest-dependent species would have strong positive correlations with this variable (cf. McGarigal and McComb 1995; Saab 1999; Donovan and Flather 2002; Haslem and Bennett 2008; Fortin and Melles 2009). These former correlations, though perhaps providing information about scale effects, are not particularly useful to resource managers because they suggest only where species do *not* occur, not which other successional stages and what structural aspects of those stages are important to each species (cf. Donovan and Flather 2002).

Guilds (Researcher Defined Assemblage Subsets)

Guild-level analyses (sensu a researcher defined taxonomic subset of interest; see Morrison and Hall 2002) represent a third approach and, depending on the objectives of the study, may have advantages over both individual species and full assemblage-level analyses. Wiens (1989a) suggested that a guild-level approach allows development of more predictive models and adds a greater degree of formality and defensibility to management decisions. We agree. First, even when researcher defined, the functional similarity of guild members means that the suite of variables used to characterize the solid or edge patch type (see Sect. 1.3) associated with a guild may not be that much more extensive than that used to describe the habitat-level patch type for any individual species within the guild (see shared patch types and similarity of variables for the sapling–shrub–opening-associated HOWR, INBU, and NAWA in Tables 6.8 and 6.9). Hence, although insight is gained into the habitat relationships of multiple species, there is an economy in collection of the explanatory data set. Inferences from the data analysis at this level also allow extrapolation to management prescriptions for a wider range of species (cf. recommendations in Hallworth et al. 2008 versus Keller et al. 2003).

Second, in comparison to a broader assemblage-level analysis, the functional grouping of species into guilds allows selection of a subset of local and biotope-scale variables that more precisely describe the particular patch type associated with each guild identified in the analysis (Holmes et al. 1979). This is in contrast to broader range descriptors typically used to characterize an entire plant community or landscape (full environmental gradient). As can be seen in the canonical multivariate-space depictions of many analyses, clusters of species, frequently identifiable as guilds, are associated with certain variable subsets and not others (e.g., MacFaden and Capen 2002; Allen et al. 2006). By focusing on functionally more precise descriptors of the individual solid or edge patch types used by guilds, it should be possible to achieve a higher level of explanatory power for species occurrences and species richness with an economy of explanatory variables (cf. Rehm and Baldassarre 2007) than using more broad-brush approaches. If estimates of total assemblage richness are the study objective, they can then be developed by summing the richness of individual guilds within the community or landscape.

Appendix C

Landscape component classification system for the Connecticut Hill WMA. Percentages are the proportion of a single GIS map cell (actual area $= 100$ m^2) represented by the type. See text for explanation of air photo analysis.

Component type	Description
1	>85 % sprouts[a] or shrubs
2	10–85 % sprouts or shrubs
3	10–85 % sprout/sapling conifers
4	<10 % sprouts or shrubs (i.e., open grass)
5	Deciduous sapling/pole w/>33 % sprouts
6	Deciduous sapling/pole w/<33 % sprouts and >67 % canopy closure
7	Deciduous sawtimber w/>33 % sprouts
8	Deciduous sawtimber w/<33 % sprouts
9	Conifer pole/sawtimber w/>33 % sprouts or w/live branches to ground level
10	Conifer pole/sawtimber w/<33 % sprouts or w/dead branches to ground level
11	Fruit trees or tall (>3 m) shrubs
12	Bare ground
13	Water
14	Mixed deciduous–coniferous sapling/poletimber
15	Mixed deciduous–coniferous sawtimber
16	Deciduous sapling/poletimber w/<33 % sprouts and <67 % canopy closure

[a] Tree diameter classes
Sprout: <4 cm dbh (included seedlings, root suckers, and stump sprouts)
Sapling: 4–10 cm dbh
Poletimber: 10–20 cm dbh
Sawtimber: >20 cm dbh

© The Author(s) 2014
J.K. Keller and C.R. Smith, *Improving GIS-based Wildlife-Habitat Analysis*,
SpringerBriefs in Ecology, DOI 10.1007/978-3-319-09608-7

Appendix D

Definition of exploratory patch types using the landscape component classification system described in Appendix C. Types 1 and 4 are "solid" patches (see Working Definitions in Sect. 1.1). Types 9, 10, 12, 13, and 14 are "edge" patches and are identified in the description by the "/" between adjacent general component types.

© The Author(s) 2014
J.K. Keller and C.R. Smith, *Improving GIS-based Wildlife-Habitat Analysis*,
SpringerBriefs in Ecology, DOI 10.1007/978-3-319-09608-7

Patch number[a]	Patch description	Landscape components[b] composing the patch type	Vertical profile[c]
1	Deciduous dense shrubs	1, 5, 7, 9 (w/deciduous understory)	0–3 m
4	Open shrub	2, 3, 11, 16	0–3 m
9	Shrub/ opening	1 1 1 1 2 2 2 2 3 3 4 4 5[d] 2 3 4 16 5 7 9 11 5 7 5 7 16	0–3 m
10	Sapling/ opening	2 2 3 5[e] and +1 1[f] or 1[g] 5 11 5 11 5 11 2	0–3 m
12	Interior canopy/ shrub[h]	1 1 5 5 7 7 8 7 8 8	0–3 m
13	N hard- wood– hemlock/ shrub	1 5 7 15 15 15	0–3 m
14	Shrub– sapling/ opening	1 1 1 1 1 2 2 2 2 2 2 2 4 4 4 4 5 5 2 3 4 16 13 3 4 5 6 7 11 13 16 5 6 7 11 16 13 16	0–3 m

[a] Numbering system follows Keller (1986, 1990)

[b] Numbers refer to landscape component numbers in Appendix C

[c] Portion of vertical profile from which leaf area estimates were used to calculate the variables MDCLA, DEACLA, and EIDCLA (Table 6.2)

[d] Each 2(row) × N(column) matrix represents all the combinations of adjacent landscape components composing an edge patch type. For example, 1 above 2 means dense shrubs (component type 1) adjacent to open shrubs (component type 2)

[e] Used for jack pine clear-cuts (CCA-4, 8, 9) in 1977–78

[f] Used for all CCA plots other than jack pine clear-cuts in 1977–78, for all CCC and OF plots in all years, and for CCA-4, 8, 9 in 1979–80

[g] Used for all CCA plots in 1979–81 except CCA-4, 8, 9 and for both CCB plots in all years

[h] Measurements of the patch type (e.g., using ESCAN) were made only on internal edges (i.e., not including the plot border)

Appendix E

Variables (Table 6.2) used in MAXR regression analysis of species density and occurrence for seven species of breeding birds at the Connecticut Hill WMA. Numbers in the table refer to patch types described in Appendix D. Explanatory variables included in the analysis for each species were selected based on examination of a correlation matrix of all of the original variables, preliminary regression analyses, and the objective of limiting the total number of variables considered to not more than 15 % of the total number of dependent data points.

Variable	Species[a]						
	ALFL	HOWR	NAWA	CSWA	BAWW	BTBW	INBU
DEAC[b]	10[c#]				13	13	14#
DEAC/PS					13*		
DEACLA	10#	14#			12,13	13	
DEACLA/PS	10*	9*	9*		13*	13*	10*
DEACLALowMid					13		
DEACLAMID	9*		9#				
DEACLAratio	10#	9#, 10#	9				
DEACLAratio/PS	10*						9*, 10*
DIST	10					13	
EIDCLA	10	9#	9#		13	13	9*
EIDCLA/PS	10*		14*				
EIDCLAratio		10*	9#, 10#				
EIDEAC	10		9#		13#	13	
LA1M				+	+	+	+#
LA1Mratio	+	+	+#				+#

(continued)

© The Author(s) 2014
J.K. Keller and C.R. Smith, *Improving GIS-based Wildlife-Habitat Analysis*,
SpringerBriefs in Ecology, DOI 10.1007/978-3-319-09608-7

(continued)

Variable	Species[a]						
	ALFL	HOWR	NAWA	CSWA	BAWW	BTBW	INBU
LALOW				+	+	+	+[#]
LALOWratio		+	+[*]				+
LALOWMID				+	+		
LALOWMIDratio		+[*]					
LAMID		+	+[*]				+
LAMIDratio	+	+	+[*]				+
MDC			4[#]	1[#]			
MDC/PS			4[*]	1*			4[*]
MDCLA		4[#]		1[#]			
MDCLA/PS		4[*]	4[*]	1*			4[*]
MDCLAMID			4				
MDCLAratio		4[#]					
MDCLAratio/PS							4[*]
NUMHAB	10	9	9, 14[#]		13	13	9
PLOTEI	10	10	10[*], 14[#]		12, 13	13	9[#]
VMRI	10[#]	9[*], 12[*]	10	1	13	13	14[*]

[a] Species abbreviations as in Table 6.1
[b] Variable definitions are found in Table 6.2
[c] Numbers refer to patch types in Table 6.1 and Appendix D
[*] Variable used in regression of density only
[#] Variable used in regression of occurrence only

References

Aber, J.C. 1979. A method for estimating foliage-height profiles in broad-leaved forests. Journal of Ecology 67:35-40.

Addicott, J.F., J.M. Aho, M.F. Antolin, D.K. Padilla, J.S. Richardson, and D.A. Soluk. 1987. Ecological neighborhoods: Scaling environmental patterns. Oikos 49: 340-346.

Aebischer, N.J., P.A. Robertson, R.E. Kenward. 1993. Compositional analysis of habitat from animal radio-tracking data. Ecology 74:1313-1325.

Ahlering, M.A. and J. Faaborg. 2006. Avian habitat management meets conspecific attraction: If you build it, will they come? Auk 123: 301-313.

Allen, J.C, S.M. Krieger, J.R. Walters, and J.A. Collazo. 2006. Associations of breeding birds with fire-influenced and riparian-upland gradients in a longleaf pine ecosystem. Auk 123: 1110-1128.

Allen, T.F.H., and T.W. Hoekstra. 1992. Toward a Unified Ecology. Columbia University Press, New York, New York. USA.

Andersen, D.E., and O.J. Rongstad. 1989. Home-range estimates of Red-tailed Hawks based on random and systematic relocations. Journal of Wildlife Management 53:802-807.

Anderson, J.R., E.E. Hardy, J.T. Roach, and R.E. Wittmer. 1976. A land use and land cover classification system for use with remote sensor data. Geological Survey Professional Paper 964, U.S. Geological Survey, Washington, DC.

Andren, H. 1994. Effects of habitat fragmentation on birds and mammals in landscapes with different proportions of suitable habitat: a review. Oikos 71:355-366.

Andren, H. 1996. Population responses to habitat fragmentation: statistical power and the random sample hypothesis. Oikos 76:235-242.

Angel, S., J. Parent and D.L. Civco. 2010. Ten compactness properties of circles: measuring shape in geography. Canadian Geographer 54:441-461.

Anich, N.M., J.A. Trick, K.M. Grveles, and J.L. Goyette. 2011. Characteristics of a red pine plantation occupied by Kirtland's warblers in Wisconsin. Wilson Journal of Ornithology 123:199-205.

Arponen, A., J. Lehtomäki, J. Lepännen, E. Tomppo, and A. Moilanen. 2012. Effects of connectivity and spatial resolution of analyses on conservation prioritization across large extents. Conservation Biology 26:294-304.

Avery, T.E., and G.L. Berlin. 1985. Fundamentals of Remote Sensing and Airphoto Interpretation. Macmillan. New York.

Bakermans, M.H. and A.D. Rodewald. 2006. Scale-dependent habitat use of Acadian flycatcher (*Empidonax virescens*) in central Ohio. Auk 123:368-382.

Bakermans, M.H. and A.D. Rodewald. 2009. Think globally, manage locally: the importance of steady-state forest features for a declining songbird. Forest Ecology and Management 258:224–232.

© The Author(s) 2014
J.K. Keller and C.R. Smith, *Improving GIS-based Wildlife-Habitat Analysis*,
SpringerBriefs in Ecology, DOI 10.1007/978-3-319-09608-7

Balbontin, J. 2005. Identifying suitable habitat for dispersal in Bonelli's eagle: An important issue in halting its decline in Europe. Biological Conservation 126:74-83.

Barg, J.J., J. Jones, and R.J. Robertson. 2005. Describing breeding territories of migratory passerines: suggestions for sampling, choice of estimator, and delineation of core areas. Journal of Animal Ecology 74:139-149.

Barg, J.J., D.V. Aiama, J. Jones and R.J. Robertson. 2006. Within-territory habitat use and microhabitat selection by male cerulean warblers (*Dendroica cerulean*). Auk 123:795-806.

Bart, J., M. Hofschen, and B.G. Peterjohn. 1995. Reliability of the Breeding Bird Survey: effects of restricting surveys to roads. Auk 112:758-761.

Bedard, J. and G. laPointe. 1984. Banding returns, arrival times, and site fidelity in Savannah Sparrows. Wilson Bulletin 96:196-205.

Bellis, L.M., A.M. Pidgeon, V.C. Radeloff, V. St.-Louis, J.L. Navarro, and M.B. Martella. 2008. Modeling habitat suitability for Greater Rheas based on satellite image texture. Ecological Applications 18:1956-1966.

Betts, M.G., A.W. Diamond, G.J. Forbes, M.-A. Villard, and J.S. Gunn. 2006. The importance of spatial autocorrelation, extent and resolution in predicting forest bird occurrence. Ecological Modeling 191:197-224.

Betts, M.G., G.J. Forbes, and A.W. Diamond. 2007. Thresholds in songbird occurrence in relation to landscape structure. Conservation Biology 21:1046-1058.

Betts, M.G., S.E. Franklin, and R.G. Taylor. 2003. Interpretation of landscape pattern and habitat change for local indicator species using satellite imagery and geographic information system data in New Brunswick, Canada. Canadian Journal of Forest Research 33:1821-1831.

Betts, M.G., J.C. Hagar, J.W. Rivers, J.D. Alexander, K. McGarigal, and B.C. McComb. 2010. Thresholds of forest bird occurrence as a function of the amount of early-seral broadleaf forest at landscape scales. Ecological Applications 20:2116-2130.

Blake, J.G. and Hoppes, W.G. 1986. Influence of resource abundance on use of tree-fall gaps by birds in an isolated woodlot. Auk 103:328-340.

Block, W.M. and L.A. Brennan. 1993. The habitat concept in ornithology: Theory and applications. Current Ornithology 11, (D. M. Power, Ed.). Plenum Press, New York.

Bock M., P. Xofis, J. Mitchley, G. Rossner, and M. Wissen. 2005. Object-oriented methods for habitat mapping at multiple scales - Case studies from Northern Germany and Wye Downs, UK. Journal for Nature Conservation 13:75-89.

Bohm, S.M. and E.K.V. Kalko. 2009 Patterns of resource use in an assemblage of birds in the canopy of a temperate alluvial forest. Journal of Ornithology 150:799-814.

Boisbunon, A., S. Canu, D. Fourdrinier, W. Strawderman, and M.T. Wells. 2013. AIC and Cp as estimators of loss for spherically symmetric distributions. arXiv:1308.2766.

Bollinger, E. K. 1995. Successional changes and habitat selection in hayfield bird communities. Auk 112:720-730.

Bollinger, E. K., and T. A. Gavin. 2004. Responses of nesting bobolinks (Dolichonyx oryzivorus) to habitat edges. Auk 121:767-776.

Boves, T.J., D.A. Buehler, J. Sheehan, P.B. Wood, A.D. Rodewald, J.L. Larken, P.D. Keyser, F.L. Newell, A. Evans, G.A. George, and T.B. Wigley. 2013. Spatial variation in breeding habitat selection by Cerulean Warblers (*Setophaga cerulea*) throughout the Appalachian Mountains. Auk 130:46–59.

Box, G.E.P., and W.J. Hill. 1967. Discrimination among mechanistic models. Technometrics 9:57-71.

Bradshaw, G.A, and M.-J.Fortin. 2000. Landscape heterogeneity effects on scaling and monitoring large scale areas using remote sensing data. Geographic Information Sciences 6:61-68.

Brennan, S.P. and G.D. Schnell. 2007 Multi-scale analysis of Tyrannid abundances and landscape variables in the Central Plains, USA. Wilson Journal of Ornithology 119:631-647.

Brown J.H. 1973. Species diversity of seed-eating desert rodents in sand dune habitats. Ecology 54:775-787.

Brown, J. H. 1995. Macroecology. University of Chicago Press.

Bullock, L.P., and D.A. Buehler. 2006. Avian use of early successional habitats: are regenerating forests, utility right-of-ways and reclaimed surface mines the same? Forest Ecology and Management 236:76-84.

Bump, G., R.W. Darrow, F.C. Edminster, W.F. Crissey, F. Everett. 1947. The Ruffed Grouse; Life History, Propagation, Management. New York Conservation Department, Albany, New York, USA.

Burhans, D.E., and F.R. Thompson. 1999. Habitat patch size and nesting success of yellow-breasted chats. Wilson Bulletin 111:210-215.

Burnham, K.P., and D.R. Anderson. 2002. Model Selection and Multimodel Inference: A Practical Information-Theoretic Approach. Second edition. Springer-Verlag, New York, New York, USA.

Burt, P.J. 1980. Tree pyramid structures for coding Hexagonally Sampled Binary Images. Computer Graphics and Image Processing 14:271-280.

Cale, P.G., and R.J. Hobbs. 1994. Landscape heterogeneity indices: problems of scale and applicability, with particular references to animal habitat description. Pacific Conservation Biology 1:183-193.

Campbell, J.B. and R.H. Wynne. 2011. Introduction to Remote Sensing. 5th Ed. The Guilford Press. New York. 667p.

Campomizzi, A.J., J.A. Butcher, S.L. Farrell, A.G. Snelgrove, B.A. Collier, K.J. Gutzwiller, M.L. Morrison, R.N. Wilkins. 2008. Journal of Wildlife Management. 72:331-336.

Capen, D.E., J.W. Fenwick, D.B. Inkley, and A.C. Boynton. 1986. Multivariate models of songbird habitat in New England forests. Pages 171-176 in Wildlife Ecology 2000 (J. Verner, M.L. Morrison and C.J. Ralph, Eds.). University of Wisconsin Press, Madison, WI, USA.

Carpenter, J.P., Y. Wang, C. Schweitzer, and P.B. Hamel. 2011. Avian community microhabitat associations of cerulean warblers in Alabama. Wilson Journal of Ornithology 123:206-217.

Chalfoun, A.D., F.R. Thompson, III, and M.J. Ratnaswamy. 2002. Nest predators and fragmentation: a review and meta-analysis. Conservation Biology 16:306–318.

Chandler, R.B., D.I. King, and S. DeStefano. 2009. Scrub-shrub bird habitat associations at multiple scales in beaver meadows in Massachusetts. Auk 126:186-197.

Chapa–Vargas, L., and S.K. Robinson. 2007. Nesting success of Acadian flycatchers (*Empidonax virescens*) in floodplain forest corridors. Auk 124:1267-1280.

Clements, F.E. 1905. Research Methods in Ecology. University Publishing Company. Lincoln, Nebraska, USA.

Cody, M.L. 1981. Habitat selection in birds: the roles of vegetation structure, competitors, and productivity. Bioscience 31:107-113.

Collier, B.A., S.L. Farrell, A.M. Long, A.J. Campomizzi, K.B. Hays, J.L. Laake, M.L. Morrison, and R.N. Wilkins. 2013. Modeling spatially explicit densities of endangered avian species in a heterogeneous landscape. Auk 13:666−676.

Collier, B.A., J.E. Groce, M.L. Morrison, J.C. Newnam, A.J. Campomizzi, S.L. Farrell, H.A. Mathewson, R.T. Snelgrove, R.J. Carroll, and R.N. Wilkins. 2012. Predicting patch occupancy in fragmented landscapes at the rangewide scale for an endangered species: an example of an American warbler. Diversity and Distributions 18:158-167.

Collins, B.M., C.K. Williams, and P.M. Castelli. 2009. Reproduction and microhabitat selection in a sharply declining Northern Bobwhite population. Wilson Journal of Ornithology 121:688-695.

Collins, S.L. and S.M. Glenn. 1997. Effects of organismal and distance scaling on analysis of species distribution and abundance. Ecological Applications 7(2):543-551.

Confer, J.L., and K. Knapp. 1979. The changing proportion of blue-winged and golden-winged warblers in Tompkins County and their habitat selection. The Kingbird 29:8-14.

Confer, J. L. and S. M. Pascoe. 2003. The avian community on utility rights-of-ways and other managed shrublands in northeastern United States. Forest Ecology and Management 85:193-206.

Congalton R.G., and K. Green. 1998. Assessing the Accuracy of Remotely Sensed Data: Principles and Practices. CRC Press, Boca Raton, Florida, USA.

Conner, E.F. and D.S. Simberloff. 1979. The assembly of species communities: chance or competition? Ecology 60:1132-1140.

Conner, R.N. and C.S. Adkisson. 1975. Effects of clearcutting on the diversity of breeding birds. Journal of Forestry 73:781-785.

Cornell, K.L. and T.M. Donovan. 2010. Scale-dependent mechanisms of habitat selection for a migratory passerine: an experimental approach. Auk 127: 899-908.

Cottam, G. and J.T. Curtis. 1956. The use of distance measures in phytosociological sampling. Ecology 37:451–460.

Covich, A.P. 1976. Analyzing shapes of foraging areas: some ecological and economical theories. Annual Review of Ecology and Systematics 7:235-258.

Cowardin, L.M., Carter, V., Golet, F.C. and E.T. LaRoe. 1979. Classification of wetlands and deepwater habitats of the United States. FWS/OBS-79/31. USFWS Office of Biological Services, US Dept. of Interior, Washington, D.C. 131 pp.

Cressie, N.A.C. 1993. Statistics for Spatial Data. Wiley, New York, NY.

Cronin, J.T. 2009. Habitat edges, within-patch dispersion of hosts, and parasitoid oviposition behavior. Ecology 90:196-207.

Cushman, S.A., and K. McGarigal. 2002. Hierarchical, multi-scale decomposition of species-environment relationships. Landscape Ecology 17:637-646.

Cushman, S.A., and K. McGarigal. 2003. Landscape-level patterns of avian diversity in the Oregon Coast Range. Ecological Monographs 73:259-281.

Cushman, S.A., and K. McGarigal. 2004. Patterns in the species-environment relationship depend on both scale and choice of response variables. Oikos 105:117-124.

Dale, M.R.T., P. Dixon, M.-J. Fortin, P. Legendre, D.E. Myers and M.S. Rosenberg. 2002. Conceptual and mathematical relationships among methods for spatial analysis. Ecography 25:558-577.

Darveau, M., P. Beauchesne, L. Belanger, J. Huot, and P. LaRue. 1995. Riparian forest strips as habitat for breeding birds in boreal forest. Journal of Wildlife Management 59:67-78.

Davis, F.W., C. Seo, and W.J. Zielinski. 2007. Regional variation in home-range-scale habitat models for fisher (*Martes pennanti*) in California. Ecological Applications 17:2195-2213.

Davis, S.K. 2004. Area sensitivity in grassland passerines: effects of patch size, patch shape, and vegetation structure on bird abundance and occurrence in southern Saskatchewan. Auk 121: 1130–1145.

deCalesta, D.S. 1994. Effect of white-tailed deer on songbirds within managed forests in Pennsylvania. Journal of Wildlife Management 58:711-718.

Dechant, J.A., M.L. Sondreal, D.H. Johnson, L.D. Igl, C.M. Goldade, A.L. Zimmerman, and B.R. Euliss. 2003. Effects of management practices on grassland birds: Bobolink. Northern Prairie Wildlife Research Center, Jamestown, ND. Northern Prairie Wildlife Research Center Online.

DeGraaf, R.M. 1991. Breeding bird assemblages in managed northern hardwood forests in New England. Pages 153-171 *in* Wildlife and Habitats in Managed Landscapes (J.E. Rodiek and E.G. Bolen, Eds.). Island Press, Washington, D.C.

DeGraaf, R.M., and R.I. Miller. 1996. The importance of disturbance and land-use history in New England: implications for forested landscapes and wildlife conservation. Pages 3-35 *in* Conservation of Faunal Diversity in Forested Landscapes (R.M. DeGraaf and R.I. Miller, Eds.). Chapman and Hall, New York.

DeGraaf, R.M., J.B. Hestbeck, and M. Yamasaki. 1998. Association between breeding bird abundance and stand structure in the White Mountains, New Hampshire and Maine, USA. Forest Ecology and Management 103 217-233.

de Knegt, H.J., F. Van Langevelde, M.B. Coughenour, A.K. Skidmore, W.F. De Boer, I.M.A. Heitkonig, N.M. Knox, R. Slotow, C. Van Der Waal, and H.H.T. Prins. 2010 Spatial autocorrelation and scaling of species-environment relationships. Ecology 91:2455-2465.

Deppe, J.L., and J.T. Rotenberry. 2008. Scale-dependent habitat use by fall migratory birds: vegetation architecture, floristics, and geographic consistency. Ecological Monographs 78:461-487.

Dettmers, R. 2003. Status and conservation of shrubland birds in the northeastern US. Forest Ecology and Management 185:81-93.

Dettmers, R., and J. Bart. 1999. A GIS modeling method applied to predicting songbird habitat. Ecological Applications 9 152-163.

Dickson, B.G., B.R. Noon, C.H. Flather, S. Jentsch, and W.R. Block. 2009. Quantifying the multi-scale response of avifauna to prescribed fire experiments in the southwest United States. Ecological Applications 19:608-621.

Diniz-Filho, J.A.F., L.M. Bini, and B.A. Hawkins. 2003. Spatial autocorrelation and red herrings in ecology. Global Ecology & Biogeography 12:53-64.

Diniz-Filho, J.A.F., B.A. Hawkins, L.M. Bini, P. De Marco, Jr., and T.M. Blackburn. 2007. Are spatial regression methods a panacea or Pandora's box? A reply to Beale et al. (2007). Ecography 30:848-851.

Diamond, J.M. and M.E. Gilpin, 1982. Examination of the "Null" model of Connor and Simberloff for species co-occurrences on islands. Oecologia 52:64-74.

Donner, D.M., J.R. Probst, and C.A. Ribic. 2008. Influence of habitat amount, arrangement, and use on population trend estimates of male Kirtland's warblers. Landscape Ecology 23:467-480.

Donovan, T. M., and C. Flather. 2002. Relationships between North American songbird trends, habitat fragmentation, and landscape occupancy. Ecological Applications 12:364–374.

Donovan, T.M., P.W. Jones, E.M. Annand, and F.R. Thompson, III. 1997. Variation in local scale edge effects: mechanisms and landscape context. Ecology 78:2064–2075.

Donovan, T.M., G.S. Warrington, W.S. Scwenk, J.H. Dinitz. 2012. Estimating landscape carrying capacity through maximum clique analysis. Ecological Applications 22:2265-2276.

Drapeau, P., A. Leduc, J-F. Giroux, J-P.L. Savard, Y. Bergeron, and W.L. Vickery. 2000. Landscape-scale disturbances and changes in bird communities of boreal mixed-wood forests. Ecological Monographs 70:423-444.

The Encyclopedia of Optical Engineering, Vol. 2 and 3. 2003. R. G. Driggers, Ed. Marcel Dekker, Inc., New York, NY.

Driscoll, M.J.L., and T.M. Donovan. 2004. Landscape context moderates edge effects: nesting success of wood thrushes in central New York. Conservation Biology 18:1330–1338.

Dueser, R.D., and H.H. Shughart, Jr., 1978. Microhabitat in a forest floor small mammal fauna. Ecology 59:89-98.

Dunning, J.B., B.J. Danielson, and H.R. Pulliam. 1992. Ecological processes that affect populations in complex landscapes. Oikos 65:169-175.

Dussault, C., R. Courtois, J. Huot, and J.-P. Ouellet. 2001. The use of forest maps for description of wildlife habitats: limits and recommendations. Canadian Journal of Forest Research 31:1227-1234.

Ernoult, A., Y. Tremauville, D. Cellier, P. Margerie, E. Langlois, and D. Alard. 2006. Potential landscape drivers of biodiversity components in a flood plain: Past or present patterns. Biological Conservation 127:1-17.

Ewers, R.W., S. Thorpe, and R. K. Didham. 2007. Synergistic interactions between edge and area effects in a heavily fragmented landscape. Ecology 88:96-106.

Eyre, F.H., Ed. 1980. Forest Cover Types of the United States and Canada. Society of American Foresters, Washington, D.C., USA.

Fahrig, L. 1997. Relative effects of habitat loss and fragmentation on population extinction. Journal of Wildlife Management 61:603-610.

Fahrig, L. 1998. When does fragmentation of breeding habitat affect population survival? Ecological Modelling 105:273-292.

Farrell, S.L., B.A. Collier, K.L. Skow, A.M. Long, A.J. Campomizzi, M.L. Morrison, K.B. Hays, and R.N. Wilkins. 2013. Using LiDAR-derived vegetation metrics for high-resolution species distribution models for conservation planning. Ecosphere 4(3):42. http://dx.doi.org/10.1890/ES12-000352.1.

Federal Geographic Data Committee Vegetation Subcommittee. 2008. National Vegetation Classification Standard, Version 2. FGDC-STD-005-2008 (Version 2). U.S. Department of the Interior, Geological Survey, Reston, VA. (http://www.fgdc.gov/standards/projects/FGDC-standards-projects/vegetation/NVCS_V2_FINAL_2008-02.pdf/)

Fischer, J., D.B. Lindenmayer, and I. Fazey. 2004. Appreciating ecological complexity: habitat contours as a conceptual landscape model. Conservation Biology, 18:1245–1253.

Flather, C.H., and M. Bevers. 2002. Patchy reaction-diffusion and population abundance: The relative importance of habitat amount and arrangement. American Naturalist 159:40-56.

Fleming, K.K., K.A. Didier, B.R. Miranda, and W.F. Porter. 2004. Sensitivity of a white-tailed deer habitat-suitability index model to error in satellite land-cover data: implications for wildlife habitat-suitability studies. Wildlife Society Bulletin 32:158-168.

Forman, R.T. 1995. Land Mosaics: the Ecology of Landscapes and Regions. Cambridge University Press, New York, New York, USA.

Forman, R.T., A.E. Galli, and C. Leck. 1976. Forest size and avian diversity in New Jersey woodlots with some land use implications. Oecologia 26:1-8.

Forman, R.T., and M. Godron. 1981. Patches and structural components for a landscape ecology. Bioscience 31:733-740.

Forman, R.T.T., and M. Godron. 1986. Landscape Ecology. Wiley, New York, New York, USA

Fortin, M.-J., M.R.T. Dale, and J. ver Hoef. 2002. Spatial analysis in ecology. Pages 2051-2058 in Encyclopedia of Environmetrics Vol 4 (A.H. El-Shaarawi and W.W. Piegorsch, Eds.).

Fortin, M.-J., and S.J. Melles. 2009. Avian spatial responses to forest spatial heterogeneity at the landscape level: conceptual and statistical challenges. Pages 137-160 in Real World Ecology: Large-Scale and Long-Term Case Studies and Methods (S. Miao, S. Carstenn & M. Nungesser, Eds.). Springer, New York.

Franklin, J., D.K. Simons, D. Beardsley, J.M. Rogan, and H. Gordan. 2000. Evaluating errors in a digital vegetation map with forest inventory data and accuracy assessment using fuzzy sets. Transactions in Geographic Information Systems 5:285-304.

Freemark, K.E., J.B. Dunning, S.J. Hejl, and J.R. Probst. 1995. A landscape ecology perspective for research, conservation, and management. in Ecology and Management of Neotropical Migratory Birds (T.E. Martin and D.M. Finch, Eds.). Oxford University Press, New York.

Freitas, G.H.S., and M. Rodrigues. 2012. Territory distribution and habitat selection of the Serra Finch (Embernagra longicauda) in Serra do Cipo, Brazil. Wilson Journal of Ornithology 124:57-65.

Fretwell, S.D., and H.L. Lucas. 1970. On territorial behavior and other factors influencing habitat distribution in birds. I. Theoretical development. Acta Biotheoretica 14:16-36.

Gallant, A.L. 2009. What you should know about land-cover data. Journal of Wildlife Management 73:796-805.

Galli, A.E., C. Leck, and R.T. Forman. 1976. Avian distribution patterns in forest islands of New Jersey. Auk 93:356-364.

Gaston, K.J., and T.M. Blackburn. 2000. Pattern and Processes in Macroecology. Blackwell Scientific, Oxford.

Gilbert, F.S. 1980. The equilibrium theory of island biogeography: fact or fiction? Journal of Biogeography 7:209-236.

Gilpin, M.E., and J.M. Diamond. 1982. Factors contributing to non-randomness in species co-occurrences on islands. Oecologia 52:75-84.

Glenn, E.M., and W.J. Ripple. 2004. On using digital maps to assess wildlife habitat. Wildlife Society Bulletin 32:852-860.

Goetz, S.J., D. Steinberg, M.G. Betts, R.T. Holmes, P.J. Doran, R. Dubayah, and M. Hoften. 2010. LiDAR remote sensing variables predict breeding habitat of a Neotropical migrant bird. Ecology 91:1569-1582.

Gorman, O.T. and J.R. Karr. 1978. Habitat structure and stream fish communities. Ecology 59:507-515.

Gosz, J.R. 1991. Fundamental ecological characteristics of landscape boundaries. Pages 8-30 in Ecotones: The Role of Landscape Boundaries in the Management and Restoration of Changing Environments (M.M. Holland, P.G. Risser, and R.J. Naiman, Eds.). Chapman and Hall, New York.

Gottschalk, T.K., F. Huettmann, and M. Ehlers. 2005. Thirty years of analyzing and modeling avian habitat relationships using satellite imagery data: a review. International Journal of Remote Sensing 26:2631-2656.

Graf, R.F., L. Mathys, and K. Bollman. 2009. Habitat assessment for forest dwelling species using LiDAR remote sensing: Capercaille in the Alps. Forest Ecology and Management 257:160-167.

Graham, M. 2003. Confronting multicolinearity in ecological multiple regression. Ecology 84:2809-2815.

Grant, T.A., E. Madden, and G.B. Berkey. 2004. Tree and shrub invasion in northern mixed-grass prairie: implications for breeding grassland birds. Wildlife Society Bulletin 32:807-818.

Graves, B.M., A.D. Rodewald, and S.D. Hull. 2010. Influence of woody vegetation on grassland birds within reclaimed surface mines. Wilson Journal of Ornithology 122:646-654.

Green, R.H. 1979. Sampling Design and Statistical Methods for Environmental Biologists. John Wiley and Sons, New York.

Grinnell, J. 1917. The niche-relationships of the California thrasher. Auk 34:427-433.

Grinnell, J. 1924. Geography and evolution. Ecology 5:225-229.

Grinnell, J. 1928. Presence and absence of animals. University of California Chronicles 30:429-450.

Guo, X., J. Wilmhurst, S. McCanny, P. Fargey, and P. Richard. 2004. Measuring spatial and vertical heterogeneity of grasslands using remote sensing techniques. Journal of Environmental Informatics 3:24-32.

Gustafson, E. J. 1998. Quantifying landscape spatial pattern: what is the state of the art? Ecosystems 1:143-156.

Gustafson, E.J., and G.R. Parker. 1992. Relationships between landcover proportion and indices of landscape spatial pattern. Landscape Ecology 7:101-110.

Habibzadeh, N., M. Karami, S.K. Alavipanah, and B. Riazi. 2013. Landscape requirements of Caucasian Grouse (*Lyrurus mlokosiewiczi*) in Arasbaran region, East Azerbaijan, Iran. Wilson Journal of Ornithology 125:140-149.

Hagan, J,M., P.S. McKinley, A.L. Meehan, and S.L. Grove 1997. Diversity and abundance of landbirds in an industrial forest in Maine. Journal of Wildlife Management 61:718-733.

Hagan, J.M., and A.L. Meehan. 2002. The effectiveness of stand-level and landscape-level variables for explaining bird occurrence in an industrial forest. Forest Science 48:231-242.

Hall, L.S., P.R. Krausman, and M.L. Morrison. 1997. The habitat concept and a plea for standard terminology. Wildlife Society Bulletin 25:173-182.

Hallworth, M., A. Ueland, E. Anderson, J.D. Lambert, and L. Rietsma. 2008. Habitat selection and site fidelity of Canada Warblers (*Wilsonia canadensis*) in central New Hampshire. Auk 125:880-888.

Hanson, H., 1962. Dictionary of Ecology. Peter Owen, London.

Hartman, P.J., D.S. Maehr, and J.L. Larkin. 2009. Habitat selection by Cerulean Warblers in Eastern Kentucky. Wilson Journal of Ornithology 121:469-475.

Harvey, P.H., P.J. Greenwood and C.M. Perrins. 1979. Breeding area fidelity of Great Tits (*Parus major*). Journal of Animal Ecology 48:305-313.

Haslem, A., and A.F. Bennett. 2008. Birds in agricultural mosaics: the influence of landscape pattern and countryside heterogeneity. Ecological Applications 18:185-196.

Heglund, P.J. 2002. Foundations of species-environment relations. Pages 35-41 *in* Predicting Species Occurrences: Issues of Accuracy and Scale (J.M. Scott, P.J. Heglund, and M.L. Morrison, Eds.). Island Press, Washington, DC.

Hiebeler, D. 2000. Populations on fragmented landscapes with spatially structured heterogeneities: landscape generation and local dispersal. Ecology 81:1629-1641.

Hilden, O. 1965. Habitat selection in birds: a review. Annales Zoologici Fennici 2:53-75.

Hill, M.O., and H.G. Gauch. 1980. Detrended correspondence analysis: an improved ordination technique. Vegetatio 42:47–58.

Hines, E.M., J. Franklin, and J.R. Stephenson. 2005. Estimating the effects of map error on habitat delineation for the California spotted owl in Southern California. Transactions in GIS 9:541-559.

Hodges, M.F., Jr., and D.G. Krementz. 1996. Neotropical migratory breeding bird communities in riparian forests of different widths along the Altamaha River, Georgia. Wilson Bulletin 108:496-506.

Holland, J.D., D.G. Bert, and L Fahrig. 2004. Determining the spatial scale of species' response to habitat. Bioscience 54:227-233.

Holmes, R.T., R.E. Bonney, Jr. and S.W. Pacala. 1979. Guild structure of the Hubbard Brook bird community: a multivariate approach. Ecology 60:512-520.

Holmes, R. T., and T. W. Sherry. 2001. Thirty-year bird population trends in an unfragmented temperate deciduous forest: importance of habitat change. Auk 118:589-609.

Homan, R.N., C.D. Wright, G.L. White, L.F. Michael, B.S. Slaby, and S.E. Edwards. 2008. Multiyear study of the migration orientation of Ambystoma maculatum (Spotted Salamanders) among varying terrestrial habitat. Journal of Herpetology 42:600-607.

Hooper, R.G., H.S. Crawford, and R.F. Harlow. 1973. Bird density and diversity as related to vegetation in forest recreational areas. Journal of Forestry 71:766-769.

Howell, C.A., S.C. Latta, T.M. Donovan, P.A. Porneluzi, G.R. Parks, and J. Faaborg. 2000. Landscape effects mediate breeding bird abundance in Midwestern forests. Landscape Ecology 15:547-562.

Howell, J.E., J.T. Peterson, and M.J. Conroy. 2008. Building hierarchical models of avian distribution for the state of Georgia. Journal of Wildlife Management 72:168-178.

Huber, T.P., and K.E. Casler. 1990. Initial analysis of Landsat TM data for elk habitat mapping. International Journal of Remote Sensing 11:907-912.

Huston, M.A., 2002. Introductory essay: Critical issues for improving predictions. Pages 7-21 in Predicting Species Occurrences: Issues of Accuracy and Scale (J.M. Scott, P.J. Heglund, and M.L. Morrison, Eds.). Island Press, Washington, DC.

Hutchinson, G.E. 1957. Concluding remarks. Cold Spring Harbor Symposium on Quantitative Biology (Population Studies: Animal Ecology and Demography) 22:415-427.

Hutto, R.L. 2014. Time budgets of male Calliope hummingbirds on a dispersed lek. Wilson Journal of Ornithology 126:121-128.

Imbeau, L., P. Drapeau, and M. Monkkonen. 2003. Are forest birds categorized as "edge species" strictly associated with edges? Ecography 26:514-520.

Jaksic, F.M. 1981. Abuse and misuse of the term "guild" in ecological studies. Oikos 37:397-400.

James, F.C. 1971. Ordinations of habitat relationships among breeding birds. Wilson Bulletin 83:215-236.

James, F.C., and C.E. McCulloch. 2002. Predicting species presence and abundance. Pages 461-465 in Predicting Species Occurrences: Issues of Accuracy and Scale (J.M. Scott, P.J. Heglund, and M.L. Morrison, Eds.). Island Press, Washington, DC.

James F.C. and H.H. Shugart, Jr. 1970. A quantitative method of habitat description. Audubon Field Notes 24:727-736.

Jelinski, D. E., and J. Wu. 1996. The modifiable areal unit problem and implications for landscape ecology. Landscape Ecology 11:129-140.

Johnson, C.J., D.R. Seip, and M.S. Boyce. 2004. A quantitative approach to conservation planning: using resource selection functions to map distribution of mountain caribou at multiple spatial scales. Journal of Applied Ecology 41:238-251.

Johnson, C.M., L.B. Johnson, C. Richards, and V. Beasley. 2002. Predicting occurrence of amphibians: An assessment of multiple-scale models. Pages 157-170 in Predicting Species Occurrences: Issues of Accuracy and Scale (J.M. Scott, P.J. Heglund, and M.L. Morrison, Eds.). Island Press, Washington, DC.

Johnson, D.H. 1980. The comparison of usage and availability measurements for evaluating resource preference. Ecology 61:65-71

Johnson, M.D. 2007. Measuring habitat quality: a review. Condor 109:489-504.

Jongman, R.H.G., C.J.F. ter Braak, and O.F.R. van Tongeren. 1987. Data Analysis in Community and Landscape Ecology. Pudoc, Wageningen.

Karr, J.R. 1968. Habitat and avian diversity on strip-mined land in east-central Illinois. Condor 70:348-357.

Kasumovic, M.M., L.M. Ratcliffe, and P.T. Boag. 2009. Habitat fragmentation and paternity in Least Flycatchers. Wilson Journal of Ornithology 121:306-313.

Kays, R.W., M.E. Gompper, and J.C. Ray. 2008. Landscape ecology of Eastern coyotes based on large-scale estimates of abundance. Ecological Applications 18:1014-1027.

Keller, C.M.E., C.S. Robbins, and J.S. Hatfield. 1993. Avian communities in riparian forests of different widths in Maryland and Delaware. Wetlands 13:137-144.

Keller, J.K. 1980. Species composition and density of breeding birds in several habitat types on the Connecticut Hill Wildlife Management Area. MS Thesis. Cornell University, Ithaca, New York.

Keller, J.K. 1986. Predicting avian species richness by assessing guild occupancy: the minimum critical patch hypothesis. Ph.D. Dissertation. Cornell University, Ithaca, New York.

Keller, J.K. 1990. Using aerial photography to model species-habitat relationships: the importance of habitat size and shape. Pages 34-46 in Ecosystem Management: Rare Species and Significant Habitats (R.S. Mitchell, C.J. Sheviak, D.J. Leopold, Eds.). New York State Museum Bull. 471.

Keller, J.K., D. Heimbuch, and M.E. Richmond. 1979a. A method of horizontal habitat quantification for use in open canopy communities. Pages 82-88 in Proceedings of Pecora IV Symposium, Application of Remote Sensing Data to Wildlife Management. National Wildlife Federation Scientific and Technical Series 3.

Keller, J.K., D. Heimbuch, and M.E. Richmond. 1979b. Optimization of grid cell shape for the analysis of wildlife habitat. Pages 1419-1428 in Proceedings of the Thirteenth International Symposium on Remote Sensing of Environment Volume III. Environmental Research Institute of Michigan, Ann Arbor.

Keller, J.K., D. Heimbuch, and M.E. Richmond. 1980. Optimization of grid cell shape for quantification of spatial arrangement. Pages 153-162 in Remote Sensing of Earth Resources. Volume VIII (F. Shahrokhi and T. Paluden, Eds.).

Keller, J.K. and C.R. Smith 1983. Birds in a patchwork landscape. The Living Bird Quarterly 2:20-23.

Keller, J.K., M.E. Richmond, and C.R. Smith. 2003. An explanation of patterns of breeding bird species richness and density following clearcutting in Northeastern USA forests. Journal of Forest Ecology and Management 174:541-564.

Keller, M.E., and S.H. Anderson. 1992. Avian use of habitat configurations created by forest cutting in southeastern Wyoming. Condor 94:55-65.

Kendeigh, S.C. 1961. Animal Ecology. Prentice-Hall, Inc., Englewood Cliffs, New Jersey.

Kennedy, C.M., E.H. Campbell Grant, M.C. Neel, W.F. Fagan, and P.P. Marra. 2011. Landscape matrix mediates occupancy dynamics of Neotropical avian insectivores. Ecological Applications 21:1837-1850.

Kerr, J.T., and M. Ostrovsky. 2003. From space to species, ecological applications for remote sensing. Trends in Ecology and Evolution 18:299-305.

King, D.I., R.B. Chandler, S. Schlossberg, and C.C. Chandler. 2009a. Habitat use and nest success of scrub-shrub birds in wildlife and silvicultural openings in western Massachusetts, USA. Forest Ecology and Management 257:421-426.

King, D. I., R. B. Chandler, S. Schlossberg, and C C. Chandler. 2009b. Effects of width, edge and habitat on the abundance and nesting success of scrub-shrub birds in powerline corridors. Biological Conservation 142:2672–2680.

King, D.I., and R.M. DeGraaf. 2000. Bird species diversity and nesting success in mature, clearcut and shelterwood forest in Northern New Hampshire. Forest Ecology and Management 129:227-235.

King, D.I., R.M. DeGraaf, and C.R. Griffin. 2001. Productivity of early successional shrubland birds in clearcuts and groupcuts in an eastern deciduous forest. Journal of Wildlife Management 65(2):345-350.

Kirkland, G.L., Jr. 1977. Responses of small mammals to the clearcutting of northern Appalachian forests. Journal of Mammalogy 58:600-609.

Klopfer, P.H. and J.U. Ganzhorn. 1985. Habitat selection: behavioral aspects. Pages 435-453 *in* Habitat Selection in Birds (M.L. Cody, Ed.). Academic Press, New York.

Kolasa, J. 1989. Ecological systems in hierarchical perspective: breaks in community structure and other consequences. Ecology 70:36-47.

Kotliar, N.B. and J.A. Wiens. 1990. Multiple scales of patchiness and patch structure: a hierarchical framework for the study of heterogeneity. Oikos 59:253-260.

Krawchuk, M.A., and P.D. Taylor. 2003. Changing importance of habitat structure across multiple spatial scales for three insect species. Oikos 103:153-161.

Kubel, J.E. and R.H. Yahner. 2008. Quality of anthropogenic habitats for golden-winged warblers in central Pennsylvania. Wilson Journal of Ornitholgy 120:801-812.

Kuehl, R.O. 2002. Design of Experiments: Statistical Principles of Research Design and Analysis. 2nd edition. Brooks/Cole, Pacific Grove, California, USA.

Lapin, C.N., M.A. Etterson, and G.J. Niemi. 2013. Occurrence of Connecticut Warbler Increases with size of patches of coniferous forest. Condor 115:168-177.

Lawler, J.L., R.J. O'Conner, C.T. Hunsaker, K.B. Jones, T.R. Loveland, and D. White. 2004. The effects of habitat resolution on models of avian diversity and distributions: a comparison of two land cover classifications. Landscape Ecology 19:515-530.

Laymon, S.A. and J.A. Reid. 1986. Effects of grid-cell size on tests of a spotted owl HSI model. Pages 93-96 *in* Wildlife 2000 (J. Verner, M.L. Morrison and C.J. Ralph, Eds.). University of Wisconsin Press, Madison, Wisconsin, USA.

LeBrun, J.J., W.E. Thogmartin, and J.R. Miller. 2012. Evaluating the ability of regional models to predict local avian abundance. Journal of Wildlife Management 76:1177-1187.

Lee, M., L. Fahrig, K.E. Freemark, and D.J. Currie. 2002. Importance of patch scale vs landscape scale on selected forest birds. Oikos 96:110-118.

Legendre, P. 1993. Spatial autocorrelation: trouble or new paradigm. Ecology 74:1659-1673.

Legendre, P., M.R.T. Dale, M.-J. Fortin, J. Gurevitch, M. Hohn, and D. Myers. 2002. The consequences of spatial structure for the design and analysis of ecological field surveys. Ecography 25:601-615.

Lemen, C.A., and M.L. Rosenzweig. 1978. Microhabitat selection in two species of heteromyid rodents. Oecologia 33:127-135.

Leonard, T.D., P.D. Taylor, and I.G. Warkentin. 2008. Landscape structure and spatial scale affect space use by songbirds in naturally patchy and harvested boreal forests. Condor 110:467-481.

Leopold, A. 1933. Game Management. Charles Scribner's Sons. New York.

Lichstein, J.W., T.R. Simons, and K.E. Franzreb. 2002a. Landscape effects on breeding songbird abundance in managed forests. Ecological Applications 12:836-857.

Lichstein, J.W., T.R. Simons, S.A. Shriner, and K.E. Franzreb. 2002b. Spatial autocorrelation and autoregressive models in ecology. Ecological Monographs 72:445-463.

Lillesand, T.M., R.W. Kieffer, and J.W. Chipman. 2008. Remote Sensing and Image Interpretation. 6th ed. John Wiley & Sons. New York.

Lindenmayer, D.B., Cunningham, R.B., Fischer, J., 2005. Vegetation cover thresholds and species responses. Biological Conservation 124:311–316.

Litwin, T.S. and C.R. Smith. 1992. Factors influencing the decline of Neotropical migrants in a northeastern forest fragment: Isolation, fragmentation, or mosaic effects? *In* The ecology and conservation of Neotropical migrant landbirds (J.M. Hagan and D. Johnston, Eds.). Smithsonian Institution Press, Washington, DC.

Looijen, R.C. 1998. Holism and reductionism in biology and ecology: the mutual dependence of higher and lower level research programmes. Ph.D. Dissertation. University of Groningen, Netherlands.

Loveland, T.R., J.W. Merchant, D.J. Ohlen, and J.F. Brown. 1991. Development of a land-cover characteristics database for the conterminous United States. Photogrammetric Engineering and Remote Sensing 57:1453-1463.

Luoto, M., M. Kuussaari, and T. Toivonen. 2002. Modeling butterfly distribution based on remote sensing data. Journal of Biogeography 29:1027-1037.

Luoto, M., R. Virkkala, R.H. Heikkinen, and K. Rainio. 2004. Predicting bird species richness using remote sensing in boreal agricultural-forest mosaics. Ecological Applications 14:1946-1962.

MacArthur, R.H. 1958. Population ecology of some warblers of northeastern coniferous forests. Ecology 39:599-619.

MacArthur, R.H. and J.W. MacArthur. 1961. On bird species diversity. Ecology 42:594-598.

MacArthur, R.H., J.W. MacArthur and J. Preer. 1962. On bird species diversity. II. Predictions of bird censuses from habitat measurements. American Naturalist 96:167-174.

MacArthur, R.H. and E.O. Wilson. 1967. The Theory of Island Biogeography. Princeton University Press, Princeton, New Jersey, USA.

MacFaden and Capen 2002. Avian habitat relationships at multiple scales in a New England forest. Forest Science 48:243-253.

MacKenzie, D.I., J.D. Nichols, G.B. Lachman, S. Droege, J.A. Royle, and C.A. Langtimm. 2002. Estimating site occupancy rates when detection probabilities are less than one. Ecology 83:2248–2255.

MacKenzie, D.I. and J.A. Royle. 2005. Designing occupancy studies: general advice and allocating survey effort. Journal of Applied Ecology 42:1105-1114.

Macreadie, P.I., J.S. Hindell, M.J. Keough, G.P. Jenkins, and R.M. Connolly. 2010. Resource distribution influences positive edge effects in a seagrass fish. Ecology 91:2013-2021.

Martin K.J, R.S. Lutz and M. Worland. 2007. Golden-winged Warbler habitat use and abundance in northern Wisconsin. Wilson J. Ornithology 119:523-532.

Martin, T.E. 1987. Food as a limit to breeding birds: a life history perspective. Annual Review of Ecology and Systematics 18:453-487.

Martin, T.J. 1992. Breeding productivity considerations: What are the appropriate habitat features for management? *In* Ecology and Conservation of Neotropical Migrant Landbirds (Hagan, J.M., III and D.W. Johnston, Eds.). Smithsonian Institution Press, Washington, DC. 609pp.

Matias, M.G., A.J. Underwood, and R.A. Coleman. 2007. Interactions of components of habitat alter composition and variability of assemblages. Journal of Animal Ecology 76:986-994.

Matias, M.G., A.J. Underwood, D.F. Hochuli, and R.A. Coleman. 2010. Independent effects of patch size and structural complexity on diversity of benthic macroinvertebrates. Ecology 91:1908-1915.

Maurer, B.A. 2002. Predicting distribution and abundance: Thinking within and between scales. Pages 125-132 *in* Predicting Species Occurrences: Issues of Accuracy and Scale (J.M. Scott, P.J. Heglund, and M.L. Morrison, Eds.). Island Press, Washington, DC.

Mazerolle, M.J. and M.-A. Villard. 1999. Patch characteristics and landscape context as predictors of species presence and abundance: A review. Ecoscience 6:117-124.

Mazerolle, M.J., A. Desrochers, and L. Rochefort. 2005. Landscape characteristics influence occupancy by frogs after accounting for detectability. Ecological Applications 15:824-834.

McDermid, G.J., N.C. Coops, M.A. Wulder, S.E. Franklin, and N.E. Seitz. 2010. Critical remote sensing contributions to spatial wildlife ecological knowledge and management. Pages 193-221 *in* Spatial Complexity, Informatics, and Wildlife Conservation (S.A. Cushman and F. Huettmann, Eds.). Springer, Tokyo.

McDermid, G.J., R.J. Hall, G.A. Sanchez-Azofeifa, S.E. Franklin, G.B Stenhouse, T. Kobliuk, and E.F. LeDrew. 2009. Remote sensing and forest inventory for wildlife habitat assessment. Forest Ecology and Management 257:2262-2269.

McElhone, P.M., P.B. Wood, and D.K. Dawson. 2011. Effects of stop-level habitat change on cerulean warbler detections along Breeding Bird Survey routes in the central Appalachians. Wilson Journal of Ornithology 123:699-708.

McGarigal, K. 2002. Landscape pattern metrics. Pages 1135-1142 *in* Encyclopedia of Environmetrics, Volume 2 (A.H. El-Shaarawi and W.W. Piegorsch, Eds.). John Wiley & Sons, Sussex.

McGarigal, K., S.A. Cushman, M.C. Neel, and E. Ene. 2002. FRAGSTATS v3: Spatial pattern analysis program for categorical maps. URL: www.umass.edu/landeco/research/fragstats/fragstats. html. Mladenoff, D. and B. Dezonia, (n.d.). *APACK 2.17 User's Guide*; Electronic Publication.

McGarigal, K. and S. A. Cushman. 2005. The gradient concept of landscape structure. Pages 112-119 *in* Issues and Perspectives in Landscape Ecology (J.A. Wiens and M. Moss, Eds.). Cambridge University Press, Cambridge.

McGarigal, K. and B.J. Marks. 1995. FRAGSTATS: spatial pattern analysis program for quantifying landscape structure. U.S. Department of Agriculture Forest Service, General Technical Report PNW-351, Portland, Oregon.

McGarigal, K. and W.C. McComb. 1995. Relationships between landscape structure and breeding birds in the Oregon Coast Range. Ecological Monographs 65:235-260.

McKenny, H.C., W.S. Keeton, and T.M. Donovan. 2006. Effects of structural complexity enhancement on eastern red-backed salamander (*Plethodon cinereus*) populations in northern hardwood forests. Forest Ecology and Management 230:186-196.

MacMahon, J.A., D.J. Schimpf, D.C. Anderson, K.G. Smith, and R.L. Bayn Jr. 1981. An organism-centered approach to some community and ecosystem concepts. Journal of Theoretical Biology 88:287-307.

Menard, S. 1995. Applied Logistic Regression Analysis. Quantitative Applications in the Social Sciences Series No. 07-106. Sage University, Thousand Oaks, CA.

Meyer, C.B. 2007. Does scale matter in predicting species distributions? Case study with the marbled murrelet. Ecological Applications 17:1574-1483.

Meyer, J.S., L.L. Irwin, and M.S. Boyce. 1998. Influence of habitat abundance and fragmentation on northern spotted owls in western Oregon. Wildlife Monographs 139:1-51.

Miller, R.I. 1996. Modern approaches to mapping forest diversity. Pages 595-614 *in* Conservation of Faunal Diversity in Forested Landscapes (R.M DeGraaf and R.I. Miller, Eds.). Chapman & Hall, London.

Miller, J.R., M.D. Dixon, and M.G. Turner. 2004. Response of avian communities in large-river floodplains to environmental variation at multiple scales. Ecological Applications 14:1394-1410.

Mitchell, L.R., C.R. Smith, and R.A. Malecki. 2000. Ecology of grassland breeding birds in the Northeastern United States – A literature review with recommendations for management. New York Cooperative Fish and Wildlife Research Unit, Department of Natural Resources, Cornell University, Ithaca, NY. 69pp.

Moilanen, A. and M. Nieminen. 2002. Simple connectivity measures in spatial ecology. Ecology 83:1131-1145.

Mönkkönen, M. and P. Reunanen. 1999. On critical thresholds in landscape connectivity: a management perspective. Oikos 84:302-305.

Moore, B.D., I.R. Lawler, I.R. Wallis, C.M. Beale, and W.J. Foley. 2010. Palatability mapping: a koala's eye view of spatial variation in habitat quality. Ecology 91:3165-3176.

Moorman, C.E., D.C. Guynn Jr., J.C. Kilgo. 2002. Hooded warbler nesting success adjacent to group-selection and clearcut edges in a Southeastern forest. Condor 104:366-377.

Morris, D.W. 1987. Ecological scale and habitat use. Ecology 68:362-369.

Morrison, M.L., B.G. Marcot, and R.W. Mannan 1998. Wildlife-Habitat Relationships: Concepts and Applications. 2nd ed. University of Wisconsin Press, Madison.

Morrison, M.L. and L.S. Hall. 2002. Standard terminology: Toward a common language to advance ecological understanding and application. Pages 43-52 *in* Predicting Species Occurrences: Issues of Accuracy and Scale (J.M. Scott, P.J. Heglund, and M.L. Morrison, Eds.). Island Press, Washington, DC.

Murray, L.D., and L.B. Best. 2014. Nest-site selection and reproductive success of Common Yellowthroats in managed Iowa grasslands. The Condor 116:74-83.

Murkin, H.R., E.J. Murkin, and J.P. Ball. 1997. Avian habitat selection and prairie wetland dynamics: A 10-year experiment. Ecological Applications 7:1144-1159.

Murtaugh, P.A. 2007. Simplicity and complexity in ecological data analysis. Ecology 88:56-62.

Neel, M.C., K. McGarigal and S.A. Cushman. 2004. Behavior of class-level landscape metrics across gradients of class aggregation and area. Landscape Ecology 19:435-455.

Nielsen, S.E., G.B. Stenhouse, and M.S. Boyce. 2006. A habitat-based framework for grizzly bear conservation in Alberta. Biological Conservation 130:217-229.

Nixon, C.M., L.P. Hansen, and P.A. Brewer. 1988. Characteristics of winter habitats used by deer in Illinois. Journal of Wildlife Management 52:552-555

Noss, R.F. 1991. Effects of edge and internal patchiness on avian habitat use in an old-growth Florida hammock. Natural Areas Journal 11:34-47.

O'Connell, T. J., L. E. Jackson, and R. P. Brooks. 1998. A bird community index of biotic integrity for the Mid-Atlantic Highlands. Environmental Monitoring and Assessment 51:145-156.

O'Connell, T. J., L. E. Jackson, and R. P. Brooks. 2000. Bird guilds as indicators of ecological condition in the central Appalachians. Ecological Applications 10:1706-1721.

O'Conner, R.J. 2002. The conceptual basis for species distribution modeling: Time for a paradigm shift? Pages 25-33 in Predicting Species Occurrences: Issues of Accuracy and Scale (J.M. Scott, P.J. Heglund, and M.L. Morrison, Eds.). Island Press, Washington, DC.

Odum, E.P. 1953. Fundamentals of Ecology. Saunders, Philadelphia, PA. (3rd ed.1971).

O'Neill. R.V., D.L. DeAngelis, T.F.H. Allen, and J.B. Waide. 1986. A hierarchical concept of ecosystems. Monographs in Population Biology 23. Princeton University Press, Princeton.

O'Neill, R.V., C.T. Hunsicker, S.P. Timmins, B.L. Jackson, K.B. Jones. K.H. Riitters, and J.D. Wickham. 1996. Scale problems in reporting landscape pattern at the regional scale. Landscape Ecology 11:169-180.

O'Neill, R.V., J.R. Krummel, R.H. Gardner, G. Sugihara, B. Jackson, D.L. DeAngelis, B.T. Milne, M.G. Turner, B. Zygmunt, S.W. Christensen, V.H. Dale and R.L. Graham. 1988. Indices of landscape pattern. Landscape Ecology 1:153-162.

O'Neill, R.V., S.J. Turner, V.I. Cullinan, D.P. Coffin, T. Cook, W. Conley, J. Brunt, J.M. Thomas, M.R. Conley, J. Gosz. 1991. Multiple landscape scales: An intersite comparison: Landscape Ecology 5:137-144.

Orians, G.H., and J.F. Wittenberger. 1991. Spatial and temporal scales in habitat selection. American Naturalist 137 (Supplement) S29-S49.

Orrock J.L., Pagels J.F., McShea W.J. and Harper E.K. 2000. Predicting presence and abundance of a small mammal species: the effect of scale and resolution. Ecological Applications 10:1356–1366.

Osman, R.W. 1977. The establishment and development of a marine epifaunal community. Ecological Monographs 47:37-63.

Ostapowicz, K., P. Vogt, K.H.Riitters, J. Kozak, C. Estreguil. 2008. Impact of scale on morphological spatial pattern of forest. Landscape Ecology 23:1107-1117.

Palmeirim, J.M., 1985. Using Landsat TM imagery and spatial modeling in automatic habitat evaluation and release site selection for the ruffed grouse (Galliformes: Tetraonidae). Pages 229-238 in Proceedings of the 19th International Symposium on Remote Sensing of Environment. University of Michigan, Ann Arbor.

Patton, D.R. 1975. A diversity index for quantifying habitat "edge". Wildlife Society Bulletin 3:171-173.

Pearman, P.P. 2002. The scale of community structure: Habitat variation and avian guilds in tropical forest understory. Ecological Monographs 72:19-39.

Pearson, S.M. and Gardner R.H. 1997. Neutral models: useful tools for understanding landscape patterns. Pages 215–230 in Wildlife and Landscape Ecology (J.A. Bissonette, Ed.). Springer-Verlag, New York, New York.

Penhollow, M.E., and D.F. Stauffer. 2000. Large-scale habitat relationships of Neotropical migratory birds in Virginia. Journal of Wildlife Management 64:362-373.

Perkins, K.A. 2006. Cerulean Warbler selection of forest canopy gaps. MS Thesis. West Virginia University, Morgantown.

Perkins, K.A., and P.B. Wood. 2014. Selection of forest canopy gaps by male Cerulean Warblers in West Virginia. Wilson Journal of Ornithology 126:288–297.

Pianka, E.R. 1967. On lizard species diversity: North American flatland deserts. Ecology 48:333-351.

Pickett, S.T.A., and P.S. White (Eds.) 1985. The Ecology of Natural Disturbance and Patch Dynamics. Academic Press, New York.

Pillsbury, F.C. and J.R. Miller. 2008. Habitat and landscape characteristics underlying anuran community structure along an urban-rural gradient. Ecological Applications 18:1107-1118.

Popplewell, C., S.E. Franklin, M. Hall-Beyer, and G.B. Stenhouse. 2003. Using landscape structure to classify grizzly bear density in Alberta Yellowhead Ecosystem Bear Management
Units. Ursus 14:27-34.

Potvin, F., K. Lowell, M.J. Fortin, and L. Belanger. 2001. How to test habitat selection at the
home range scale: a resampling random windows technique. Ecoscience 8:399-406.

Pribil, S. and J. Picman. 1997. The importance of using proper methodology and spatial scale in
the study of habitat selection by birds. Canadian Journal of Zoology 75:1835-1844.

Probst, J.R. 1986. A review of factors limiting the Kirtland's warbler on its breeding grounds.
American Midland Naturalist 116:87–100

Probst, J.R. 1988. Kirtland's warbler breeding biology and habitat management. Pages 28-35 in
Integrating Forest Management for Wildlife and Fish (W. Hoekstra and J. Capp, Compilers).
USDA Forest Service, General Technical Report NC-122. St. Paul, MN.

Probst, J.R. and J.Weinrich. 1993. Relating Kirtland's warbler population to changing landscape
composition and structure. Landscape Ecology 8:257-271.

Rehm, E.M., and G.A. Baldassarre. 2007. The influence of interspersion on marsh bird abundance in New York. Wilson Journal of Ornithology 119:648-654.

Rice, J., B.W. Anderson, and R.D. Ohmart. 1984. Comparison of the importance of different habitat attributes to avian community organization. Journal of Wildlife Management 48:895-911.

Riitters, K.H., J.D. Wickam, R.V. O.Neill, K.B. Jones, E.R. Smith, J.W. Coulston, T.G.
Wade, J.H. Smith. 2002. Fragmentation of continental United States forests. Ecosystems
5:815-822.

Rioux, S., M. Belisle and J-F. Giroux. 2009. Effects of landscape structure on male density and spacing patterns in wild turkeys Meleagris gallopavo depend on winter severity. Auk 126:673-683.

Risser, P.G. 1987. Landscape ecology: state of the art. Pages 1-14 in Landscape Heterogeneity
and Disturbance (M.G. Turner, Ed.). Springer-Verlag, New York.

Rodewald, A.D., and R.H. Yahner. 2001a. Avian nesting success in forested landscapes: influence
of landscape composition, stand and nest-patch microhabitat, and biotic interactions. Auk
118:1018–1028.

Rodewald, A.D., and R.H. Yahner. 2001b. Influence of landscape composition on avian community structure and associated mechanisms. Ecology 82:3493-3504.

Root, R. 1967. The niche exploitation pattern of the Blue-gray Gnatcatcher. Ecological
Monographs 37:317-350.

Rosenberg, K.V., R.D. Ohmart, and B.W. Anderson. 1982. Community organization of riparian
breeding birds: response to an annual resource peak. Auk 99:260-274.

Roth, R.R. 1976. Spatial heterogeneity and bird species diversity. Ecology 57:773-782.

Royle, J. A.; R. M. Dorazio (2008). Hierarchical Modeling and Inference in Ecology. Elsevier.

Rudnicky, T.C. and M.L. Hunter, Jr. 1993. Reversing the fragmentation perspective: Effects of
clearcut size on bird species richness in Maine. Ecological Applications 3:357-366.

Rykken, J.J., D.E. Capen, and S.P. Mahabir. 1997. Ground beetles as indicators of land type
diversity in the Green Mountains of Vermont. Conservation Biology 11:522-530.

Saab, V. 1999. Importance of spatial scale to habitat use by breeding birds in riparian forests: A
hierarchical analysis. Ecological Applications 9:135-151.

Schill, K.L., and R.H. Yahner. 2009. Nest-site selection and nest survival of early successional
birds in central Pennsylvania. Wilson Journal of Ornithology 121:476-484.

Schlossburg and King. 2008. Are shrubland bird species edge specialists? Ecological
Applications 18(6):1325-1330.

Schlossberg and King. 2009. Postlogging succession and habitat usage of shrubland birds.
Journal of Wildlife Management 73:226-231.

Schoener, T.W. 1968. Sizes of feeding territories among birds. Ecology 49:123-141.

Schumaker, N.H. 1996. Using landscape indices to predict habitat connectivity. Ecology
77:1210-1225.

Scott, J.M., P.J. Heglund, and M.L. Morrison. (Eds). 2002. Predicting Species Occurrences:
Issues of Accuracy and Scale. Island Press, Washington, DC.

Seavy, N.E., J.H. Viers, and J.K. Wood. 2009. Riparian bird response to vegetation structure: a multiscale analysis using LiDAR measurements of canopy height. Ecological Applications 19:1848-1857.

Sherry,T.W. 1979. Competitive interactions and adaptive strategies of American Redstarts and Least Flycatchers in a northern hardwoods forest. Auk 96:265-283.

Short, H.L. and S.C. Williamson. 1986. Evaluating the structure of habitat for wildlife. Pages 97-104 in Wildlife 2000 (J. Verner, M.L. Morrison, and C.J. Ralph, Eds.). University of Wisconsin Press, Madison.

Sikes, P.J. and K.A. Arnold. 1984. Movement and mortality estimates of Cliff Swallows in Texas. Wilson Bulletin 96:419-425.

Simberloff, D.S. 1976. Experimental zoogeography of islands: effects of island size. Ecology 57: 629-641.

Simberloff, D.S. 1986. Design of nature reserves. Pages 315-338 in Wildlife Conservation Evaluation (M.B. Usher, Ed.). Chapman and Hall, Ltd. New York.

Sisk, T.D., N.M. Haddad, and P.R. Ehrlich. 1997. Bird assemblages in patchy woodlands: Modeling the effects of edge and matrix habitats. Ecological Applications 7:1170-1180.

Smith, C.R. (ed.) 1990. Handbook for Atlasing American Breeding Birds. Vermont Institute of Natural Science, Woodstock, VT.

Smith, C.R., S.D. DeGloria, M.E. Richmond, S.K. Gregory, M. Laba, S.D. Smith, J.L. Braden, E.H. Fegraus, E.A. Hill, D.E. Ogureak, and J.T. Weber. 2001. The New York Gap Analysis Project Final Report. New York Cooperative Fish and Wildlife Research Unit, Cornell University, Ithaca, NY.

Smith, J.H., S.V. Stehman, J.D. Wickham, and L. Yang. 2002. Effects of landscape characteristics on land-cover class accuracy. Remote Sensing of Environment 84:342-349.

Smith, J.H., J.D. Wickham, S.V. Stehman, and L. Yang. 2003. Impacts of patch size and land cover heterogeneity on thematic image classification accuracy. Photogrammetric Engineering and Remote Sensing 68:65-70.

Smith, K.M, W.S. Keeton, T.M. Donovan, and B. Mitchell. 2008. Stand-level forest structure and avian habitat: scale dependencies in predicting occurrence in a heterogeneous forest. Forest Science 54:36-46.

Smith, L.A., E. Nol, D.M. Burke, and K.A. Elliott. 2007. Nest site selection of Rose-breasted Grosbeaks in southern Ontario. Wilson Journal 119:151-161.

Smith T.M. and H.H. Shugart. 1987. Territory size variation in the ovenbird: the role of habitat structure. Ecology 68:695-704.

Soberon, J. 2007. Grinnellian and Eltonian niches and geographic distributions of species. Ecology Letters 10:1115-1123.

Sokal, R.R., and N.L. Oden. 1978. Spatial autocorrelation in biology: 2. Some biological implications and four applications of evolutionary and ecological interest. Biological Journal of the Linnean Society 10:229-249.

Spiegelhalter, D., J.A. Thomas, and N.G. Best. 1999. WinBUGS Version 1.2 User Manual. http://www.mrc-bsu.cam.ac.uk/bugs/references/bugs-core-papers.shtml

Star, J., and J. Estes. 1990. Geographic Information Systems: An Introduction. Prentice Hall, Englewood Cliffs, New Jersey.

Stauffer, D.F. 2002. Linking populations and habitats: Where have we been? Where are we going? Pages 53-61 in Predicting Species Occurrences: Issues of Accuracy and Scale (J.M. Scott, P.J. Heglund, and M.L. Morrison, Eds.). Island Press, Washington, DC.

Stenger, J. 1958. Food habits and available food to Ovenbirds in relation to territory size. Auk 75:335-346.

St.-Louis, V. A.M. Pidgeon, V.C. Radeloff, T.J. Hawbaker, and M.K. Clayton. 2006. High resolution image texture as a predictor of bird species richness. Remote Sensing of the Environment 105:299-312.

Stoddard, M.A. and J.P. Hayes. 2005. The influence of forest management on headwater stream amphibians at multiple spatial scales. Ecological Applications 15:811-823.

Strahler, A., C. Woodcock, and J. Smith. 1986. On the nature of models in remote sensing. Remote Sensing of Environment 20:121-139.

Strayer, D.L., M.E. Power, W.F. Fagan, S.T.A. Pickett, and J. Belnap. 2003. A classification of ecological boundaries. Bioscience 53:723-729.

Suarez, A.V., K.S. Pfennig, and S.K. Robinson. 1997. Nesting success of a disturbance dependent songbird on different kinds of edges. Conservation Biology 11:928-935.

Tabachnick, B.G. and L.S. Fidell. 1996. Using Multivariate Statistics. Harper-Collins, New York.

ter Braak, C.J.F. and I.C. Prentice. 1988. A theory of gradient analysis. Advances in ecological research: 18:271-317.

ter Braak, C.J.F., and P. Smilauer. 1998. CANOCO Reference Manual and User's Guide to Canoco for Windows: Software for Canonical Community Ordination (version 4). Microcomputer Power (Ithaca, NY USA) 352 pp.

Thompson, C.M., and K. McGarigal. 2002. The influence of research scale on bald eagle habitat selection along the lower Hudson River, New York (USA). Landscape Ecology 17:569-586.

Thompson, F.R., T.M. Donovan, R.M. DeGraaf, J. Faaborg, and S.K. Robinson. 2002. A multiscale perspective of the effects of forest fragmentation on birds in eastern forests. Pages 9-19 *in* Effects of habitat fragmentation on birds in Western landscapes: Contrasts with paradigms from the Eastern United States (T.L. George and D.S. Dobkins, Eds). Studies in Avian Biology, no. 25.

Thompson, F.R., M.B. Robbins, J.A. Fitzgerald. 2012. Landscape-level forest cover is a predictor of Cerulean Warbler abundance. Wilson Journal of Ornithology 124:721-727.

Thogmartin, W.E., A.L. Gallant, T.J. Fox, M.G. Knutson, and M.J. Suarez. 2004a. A cautionary tale regarding the use of the National Land Cover Dataset 1992. Wildlife Society Bulletin 32:970-978.

Thogmartin, W.E., J.R. Sauer, and M.G. Knutson. 2004b. A hierarchical spatial model of avian abundance with application to cerulean warblers. Ecological Applications 14:1766-1779.

Tierson, W.C., G.F. Mattfeld, R.W. Sage, Jr., and D.F. Behrend. 1985. Seasonal movements and home ranges of white-tailed deer in the Adirondacks. Journal of Wildlife Management 49:760-769.

Titterington, R.W., H.S. Crawford, and B.N. Burgason. 1979. Songbird responses to commercial clear-cutting in Maine spruce-fir forests. Journal of Wildlife Management 43:602-609.

Tomoff, C.S. 1974. Avian species diversity in desert scrub. Ecology 55:396-403.

Trani (Greif), M.K. 2002. The influence of spatial scale on landscape pattern description and wildlife habitat assessment. Pages 141-155 *in* Predicting Species Occurrences: Issues of Accuracy and Scale (J.M. Scott, P.J. Heglund, and M.L. Morrison, Eds.). Island Press, Washington, DC.

Trzcinski, M. K., L. Fahrig, and G. Merriam. 1999. Independent effects of forest cover and fragmentation on the distribution of forest breeding birds. Ecological Applications 9:586-593.

Turner, M.G. 1989. Landscape ecology: the effect of pattern and process. Annual Review of Ecology and Systematics 20:171-197.

Turner, M.G., and R.H. Gardner. 1991. Quantitative methods in landscape ecology: an introduction. Pages 3-14 *in* Quantitative Methods in Landscape Ecology (M.G. Turner and R.H. Gardner, Eds.). Springer, New York.

Turner, W., S. Spector, N. Gardiner, M. Fladeland, E. Sterling, and M. Steininger. 2003. Remote sensing for biodiversity science and conservation. Trends in Ecology and Evolution 16:306-314.

Udvardy, M.F.D. 1959. Notes on the ecological concepts of habitat, biotope and niche. Ecology 40:725-728.

Urban, D.L, O'Neil, R.V., and Shugart, H.H. Jr., 1987. Landscape ecology: A hierarchical perspective can help scientists understand spatial patterns. Bioscience 37:119-127.

Van Horne, B. 2002. Approaches to habitat modeling: The tensions between pattern and process and between specificity and generality. Pages 63-72 *in* Predicting Species Occurrences: Issues of Accuracy and Scale (J.M. Scott, P.J. Heglund, and M.L. Morrison, Eds.). Island Press, Washington, DC.

Veech, J.A., M.F. Small, and J.T. Baccus. 2012. Representativeness of land cover composition along routes of the North American Breeding Bird Survey. The Auk 129:259-267.

Verschuyl, J.P., A.J. Hansen, D.B. McWethy, R. Sallabanks, and R.L. Hutto. 2008. Is the effect of forest structure on bird diversity modified by forest productivity? Ecological Applications 18:1155-1170.

Vierling, K.T., C. Bassler, R. Brandl, L.A. Vierling, I. Wies, and J. Muller. 2011. Spinning a laser web: predicting spider distributions using LiDAR. Ecological Applications 21:577-588.

Vierling, K.T., L.A. Vierling, W.A. Gould, S. Martinuzzi, and R.M. Clawges. 2008 LiDAR: shedding new light on habitat characterization and modeling. Frontiers in Ecology and the Environment: 6:90-98.

Villard, M.-A., M. K. Trzcinski, and G. Merriam. 1999. Fragmentation effects of forest birds: relative influence of woodland cover and configuration on landscape occupancy. Conservation Biology 13:774-783.

Virkkala, R. 1991. Spatial and temporal variation in bird communities and populations in north-boreal coniferous forests: a multi-scale approach. Oikos 62:59-66.

Vogt, P. 2013 GUIDOS: tools for the assessment of pattern, connectivity, and fragmentation. Geophysical Research Abstracts Vol. 15, EGU2013-12526.

Vogt, P., K.H. Riitters, C. Estreguil. J. Kozak, T.G. Wade, J.D. Wickham. 2007. Mapping spatial patterns with morphological image processing. Landscape Ecology 22:171-177.

Wagner, H.H. and M.-J. Fortin. 2005. Spatial analysis of landscapes: concepts and statistics. Ecology 86:1975-1987.

Weakland, C.A. and P.B. Wood. 2005. Cerulean Warbler (Dendroica cerulea) microhabitat and landscape-level habitat characteristics in southern West Virginia. Auk 122:497-508.

Whittaker, R.H., 1975. Communities and Ecosystems. Macmillan & Co., New York.

Whittaker, R.H., S.A. Levin, and R.B. Root. 1973. Niche, habitat, and ecotope. American Naturalist 107:321-338.

Wiens, J.A. 1974. Habitat heterogeneity and avian community structure in North American grasslands. American Midland Naturalist 91:195-213.

Wiens, J.A. 1976. Population responses to patchy environments. Annual Review of Ecology and Systematics 7:81-120.

Wiens. J.A. 1985. Habitat selection in variable environments: shrub-steppe birds. Pages 227-251 in Habitat Selection in Birds (M.L. Cody (Ed.). Academic Press, Inc., Orlando.

Wiens, J.A. 1989a. Spatial scaling in ecology. Functional Ecology 3:385-397.

Wiens, J.A. 1989b. The Ecology of Bird Communities. Cambridge University Press, Aberdeen.

Wiens, J.A. 1994. Habitat fragmentation: island v landscape perspectives on bird conservation. Ibis 137:S97-S104.

Wiens, J.A. 2002. Predicting species occurrences: progress, problems, and prospects. Pages 739-749 in Predicting Species Occurrences: Issues of Accuracy and Scale (J.M. Scott, P.J. Heglund, and M.L. Morrison, Eds.). Island Press, Washington, DC.

Wiens, J.A. and B.T. Milne. 1989. Scaling of 'landscapes' in landscape ecology, or, landscape ecology from a beetle's perspective. Landscape Ecology 3:87–96.

Wiens, J.A, and J.T. Rotenberry, and B. Van Horne. 1987. Habitat occupancy configurations of North American shrubsteppe birds: the effects of spatial scale. Oikos 48:132-147.

Williamson, K. 1970. Birds and modern forestry. Bird Study 17:167-176.

Willis, K.J. and R.J. Whittaker. 2002. Species diversity - scale matters. Science, 295:1245–1248.

Wilson, M.D., and B.D. Watts. 2008. Landscape configuration effects on distribution and abundance of Whip-poor-wills. Wilson Journal of Ornithology 120:778-783.

Winter, M., D.H. Johnson, and J.A. Shaffer. 2005. Variability in vegetation effects on density and nesting success of grassland birds. Journal of Wildlife Management 69:185–197.

Winter, M., D.H. Johnson, J.A. Shaffer, T.M. Donovan, and W.D. Svedarsky. 2006. Patch size and landscape effects on density and nesting success of grassland birds. Journal of Wildlife Management 70:158-172.

With, K.A. 1994. Using fractal analysis to assess how species perceive landscape structure. Landscape Ecology 9:25-36.

Wolters, V., J. Bengtsson, and A. Zaitsev. 2006. Relationship among species of different taxa. Ecology 97:1886-1895.

Wu, J. 2004. Effects of changing scale on landscape pattern analysis: Scaling relations. Landscape Ecology 19:125-138.

Wu, J. and R. Hobbs (Eds.). 2006. Key Topics in Landscape Ecology. Cambridge University Press, Cambridge.

Wulder, M.A., R.J. Hall, N.C. Coops, and S.E. Franklin. 2004. High spatial resolution remotely sensed data for ecosystem characterization. Bioscience 54:511-521.

Young, J.S., and R.L. Hutto. 2002. Use of regional-scale exploratory studies to determine bird-habitat relationships. Pages 107-119 *in* Predicting Species Occurrences: Issues of Accuracy and Scale (J.M. Scott, P.J. Heglund, and M.L. Morrison, Eds.). Island Press, Washington, DC.

Zimble, D.A., D.L. Evans, G.C. Carlson, R.C. Parker, S.C. Grado, P.D. Gerard. 2003. Characterizing vertical forest structure using small-footprint airborne LiDAR. Remote Sensing of Environment 87:171-182.

Zabel, C.J., J.R. Dunk, H.B. Stauffer, L.M. Roberts, B.S. Mulder, and A. Wright. 2003 Northern spotted owl habitat models for research and management application in California (USA). Ecological Applications 13:1027-1040.